BEYOND THE MOON

BEYOND THE MOON

Paolo Maffei
Translated by D. J. K. O'Connell

The MIT Press
Cambridge, Massachusetts, and London, England

This translation of the sixth (1977) edition © 1978 by the
Massachusetts Institute of Technology
Originally published in Italy under the title
"Al Di La' Della Luna," copyright © 1973 by
Arnoldo Mondadori Editore, Milano

This book was set in VIP Helvetica by The Composing Room of Michigan, Inc. It
was printed on R + E Book and bound by Halliday Lithograph Corporation in the
United States of America.

Library of Congress Cataloging in Publication Data

Maffei, Paolo.
  Beyond the moon.

  Translation of Al di là, della luna.
  Includes index.
  1. Astronomy—Popular works. I. Title.
QB44.2.M3313   523   77-27091
ISBN 0-262-13133-1

CONTENTS

PREFACE   *vii*

TRANSLATOR'S FOREWORD   *ix*

1   FROM THE MOON TO THE SUN   *1*

2   IN THE SOLAR SYSTEM   *34*

3   THE NEARBY STARS   *99*

4   VARIABLE STARS   *127*

5   THE BIRTHPLACE OF THE STARS   *167*

6   STAR CLUSTERS   *192*

7   THE GALAXY   *211*

8   BEYOND THE GALAXY   *237*

9   AMONG THE GALAXIES   *265*

10   AT THE LIMITS OF TIME AND SPACE   *296*

CONCLUSION   *326*

APPENDIXES

A   THE DETERMINATION OF THE DISTANCES OF THE STARS   333

B   THE DISCOVERY, NOMENCLATURE, AND DISTRIBUTION OF
     VARIABLE STARS   337

C   CELESTIAL ATLASES AND CATALOGS   343

D   THE MAFFEI GALAXIES   347

E   LIFE IN THE UNIVERSE   358

     INDEX   363

PREFACE

The idea of writing this book grew out of two definite impressions, both clear and yet conflicting, immediately after the *Apollo 11* exploit. On the one hand, people of today have already developed a kind of "space awareness" by weighing the importance of the moon landing and examining life and its problems in the new light that will dominate the immediate future. On the other hand, this widespread feeling, acquired in all sorts of ways (from science fiction films and comic strips or the sight of the earth as photographed from the moon, from journalistic slogans or the memorable television transmission of 21 July 1969), is accompanied by a very limited knowledge of astronomy and completely false concepts and ideas.

There are some who think that science has eclipsed science fiction and that henceforth there is no longer room for fantasy or a sense of mystery. There are some who hold that mankind has made a great leap in space and that, even if it has not already reached the limits of the universe, at least it has advanced very far and sooner or later will finish the rest of the journey. Nothing is more erroneous than these ideas, and many others of the kind, which attribute to man's lunar explorations and other, more recent exploits in space, a value and significance altogether different from their real value.

If we wish to see these exploits in their true light, we must find out what lies *beyond the moon.* It is thus the aim of this work to offer a view of the cosmos to all of those who are reaching out to the future but who have not had the chance to devote themselves to astronomical research.

This book is not a treatise on astronomy, even though it contains much information about it, but the story of a voyage in space—a voyage that starts from the moon (no longer considered as an astronomical object) and continues as far as the greatest distances that can be reached by exploration or thought. To obtain a knowledge of the universe, we shall study the more significant objects it contains, in general, at greater and greater distances from the earth. The measure of time and space will be conveyed chiefly by the appearance of the earth and sun as seen from ever greater distances and by their disappearance when, at a certain point, they are lost among the enormous numbers of other heavenly bodies. The variety of nature outside the earth-moon system, undreamt of by the uninitiated, will be demonstrated by describing some of the panoramas that can be glimpsed in this voyage—so many open doors to contemplation and imagination.

This method has been adopted to enable the reader to fathom the essence of modern astronomy and to learn its most significant results, without being distracted and wearied by the methods used to reach them. Some of these methods, as well as some recent or still controversial developments, are explained in the appendixes, which can be read after the main text, because they are not needed in order to understand it. Appendix A, however, explains the terminology relating to the distances used in the rest of the text, among other matters, and should be read after chapter 3.

The data and the illustrations have been brought up-to-date as much as possible, so that the book may prove both attractive and useful to the educated nonspecialist while remaining within the reach of those who, not having had the opportunity or interest to pursue these studies, do not thereby forfeit the right to be fully people of their time, modern individuals who, along with millions of their peers, are building the future of mankind and feel the need to see how their own labors fit within this vaster context.

TRANSLATOR'S FOREWORD

In recent times several books have appeared that describe the achievements of modern astronomy for the general public. One might well ask what need there is for one more in this field. Paolo Maffei's book has, I believe, a distinctive quality and value of its own. It presents our knowledge of the universe beyond the earth and moon in a vivid and attractive way. It is completely authoritative, without being too technical for the intelligent reader.

That the book really does fill a need is shown by the fact that four editions of the original Italian version were issued in less than two years since it first appeared. I first translated the text from the fourth edition, but in the meantime two other Italian editions have appeared; this translation is from the sixth edition, revised and updated.

D. J. K. O'Connell

BEYOND THE MOON

1 FROM THE MOON TO THE SUN

We are on the moon. We have reached it after a four-day voyage, having abandoned that marvelous multicolored globe that we now see suspended, motionless, in a very black sky. That sphere, which seems to us so large even from a distance of four hundred thousand kilometers, is the world on which we have left all that we knew previously through direct experience or by way of indirect historical and geographical information.

It is all there: delightful landscapes and frightening phenomena; calm summer nights and fresh spring mornings; cyclones, earthquakes, wars; thousands of years of history, flourishing civilizations, and the still impressive relics of former civilizations; a very rich heritage of science, art, religion, thought; hundreds of thousands of millions of living beings, among them some thousands of millions like ourselves, with their store of joys, sorrows, and hopes.

It is indeed an enormous sphere: its circumference measures 40,000 kilometers, its weight in tons is given by the number 6 followed by twenty-one zeros. Yet this enormous and extremely heavy sphere is moving in space at a speed of 100,000 kilometers per hour—a thousand times as fast as when we drive along a highway at 100 kilometers per hour.[1] This sphere is our world, the world of human beings, who have begun to abandon it in order to reach the nearest heavenly body, the moon.

The exploit of 21 July 1969 constituted the heartening triumph of the scientific knowledge accumulated over the centuries, of today's collective technological effort, and of the courage of daring individuals throughout the ages. It has opened a new road—the "cosmic way," as journalists might call it. As is the case with all roads, however, when one starts out on this road, one cannot see the end; indeed there are few roads that lead to only one destination. Thus, it is not easy to foresee today how much of this cosmic way can be traversed by man, how long it will take, or what it will cost; above all, it is almost impossible to guess at this stage what point can be reached. But it is man's destiny to be driven constantly by the thirst for knowledge and by the spirit of adventure, that, even before it moves him physically, excites him and carries him away in thought. For that reason we

[1] Given that a tortoise covers about 100 meters in an hour, it follows that our speed on a highway compared with that of the earth in space is the same as that of a tortoise compared with our speed on a highway.

wish to foresee and prepare for those times that other generations will experience. We wish above all to know what lies beyond our satellite (which we now, in some ways, know better than the earth), at greater and greater distances, to those limits we glimpsed before leaving the earth and which, as our knowledge advances, continue to recede.

Let us therefore leave the moon as well, and, in the most fantastic and adventurous exploration that man could undertake, we shall advance in space in search of whatever awaits us beyond this barren landscape. We shall become acquainted with bodies that man may someday reach and others that, because of their nature or distance, man will never be able to land on. We shall go as far as man can plant his foot, and beyond—as far as the eye can venture and thought penetrate. And at the end of this journey we shall have a much clearer idea of what the conquest of the moon means for mankind, of what we are in the universe, of what we can become.

If we were actually to travel in space and reach a heavenly body on which we could land, we should have to choose, as the first stage in our journey, a neighboring planet, such as Venus or Mars. But our exploration is not subject to physical limitations of this kind, and thus we can turn to the heavenly body it is logical to explore first. It is the body around which the earth and moon are constantly moving, making one complete revolution each year; it is the most direct source of energy for all our activities; with its light and heat, it is the star that provides life—the sun.

THE SUN'S ENERGY

From the earth, where we are protected by the atmosphere, it is possible, on occasion, to observe the sun with the naked eye—on a foggy day, for example, or at sunrise or sunset. From the moon, however, it is impossible to look at the sun directly. Dominating the black and starry sky, it floods everything with its energy, in the form of light, heat, and other types of radiation, which arrive undiminished, not reduced or altered by any absorbing medium. Here a defect in the insulation of an astronaut's space suit would suffice to cause his death from the heat of a few minutes' exposure to the merciless rays of the sun or, if the sun were below the horizon, he would perish no less horribly from frostbite. On the moon the energy that reaches us from the sun appears in its most brutal, yet purest form.

We measured this energy before reaching the moon. The measurements were carried out on the earth's surface, and then the value obtained was corrected by means of appropriate calculations and procedures to allow for atmospheric absorption and thus to find out how much solar energy reaches the outer limits of the earth's atmosphere. The resulting value is impressive. One square meter of the earth's surface receives 1,360 watts of power from the sun (at its zenith). When the atmosphere is taken into account, the power comes to roughly 1 kilowatt. Thus on a sunny day (lasting, say, 10 hours) a modest garden measuring 200 square meters receives 2,000 kilowatt hours of energy. In the form of electrical energy the same amount would cost the proprietor some forty U.S. dollars. Thus by the end of the year, hundreds of dollars' worth of solar energy would have rained down on his garden, energy he would have put only to minimal use—to grow flowers or to get a tan during the summer. If such a small piece of land receives so much energy, we can well imagine how much falls on the whole surface of the earth. This should come as no surprise when we consider that almost all the energy we use daily is nothing but converted solar energy. Electrical energy may be produced by turbines driven by water falling through ducts, but it is the sun that is continually doing the work of raising that water from sea level to the height from which it falls, in the form of rain or snow. Coal and wood, when burned, surrender the solar energy they have stored up for millions of years or for only a few years. The energy supplied by the wind or ocean currents, that contained in our food, the working of our own muscles, the intellectual effort we are making at this very moment in following this reasoning—all have their origin principally in solar energy.

Man has recently begun to use another source of energy that is completely independent of the sun—atomic energy. Clearly the future of humanity depends on using this source; yet at present its use is very limited. The rest of the energy used still comes from the sun and is one of those free gifts, like air and water, whose immense value we do not appreciate, for the simple reason that they are indispensable and without them we should never have existed.

The fact that nearly all the energy used on our planet comes from the sun should not mislead us as to the total amount received by the earth, which is 20 trillion times greater. But we shall not dwell on this point; it is not too edifying to realize what spendthrifts we are. Rather, from the result we have obtained, we can arrive at a much more interesting discovery: we can learn how much energy the sun radiates into space.

We have seen in fact that, at the distance of the earth, the power arriving from the sun measures 1,360 watts per square meter. Since the sun emits energy in all directions, the same power received by one square meter on the earth would reach all the square meters of an ideal surface which has the sun at its center and a radius equal to the distance from the sun to the earth. This distance is known; it is equal to 149,600,000 kilometers. Thus to find the total solar power in watts, we need only calculate the surface area of a sphere of that radius in square meters and multiply the result by 1,360. The computed value is 380,000 billion billion kilowatts. Translating this high value into a form more easily grasped by the imagination, we can say that in one second the sun emits more energy than mankind has consumed in the whole of its history.

Since the area of the sun's surface is known, we need only divide the total power by that area, expressed in square centimeters, to find the power coming from a square centimeter of the surface of the sun. The result is 6.2 kilowatts. Thus a piece of the solar surface the size of a postage stamp, about 5 square centimeters, emits more energy than five-hundred 60-watt lamps, one of which normally suffices to light a room of average size.

When we know the energy emitted per unit of surface, a law of physics (Stefan's Law) enables us, given certain conditions, to arrive at the temperature. Thus we find that the surface temperature of the sun is 5,750°K (degrees Kelvin or absolute), which is much higher than any temperature that can normally be obtained on the earth except by means of thermonuclear bombs.

THE SUN'S APPEARANCE

From the distance and apparent dimensions of the sun one can determine its actual size and, by a method we shall not go into, its weight. It turns out that the sun has a diameter of 1,392,000 kilometers, more than three times the distance from the earth to the moon. The value of the mass, in tons, is 199 followed by twenty-nine zeros. Our imagination can grasp that figure more easily if we say that the sun weighs as much as 322,270 planets like the earth.

Thus we see from this introduction that the sun is an immense body from which an enormous amount of energy is constantly being released.

On earth the areas in which energy is produced are nearly always characterized by violent events. We need only consider earthquakes, volcanic eruptions, cyclones, the heat in our fireplace, or water boiling in a kettle, which must appear terrifying to the eyes of a small fly. The situation on the sun is no different. That perfectly homogeneous disk, with its well-defined outline and no traces of disturbance, as seen by the naked eye, only seems homogeneous because of the distance and the limits of our vision. Very different indeed is its appearance when viewed through an instrument.

Let us begin by examining that part of the sun visible to the naked eye or through a normal telescope: the photosphere. Observed through a good telescope, it is seen to be not homogeneous but rather composed of a myriad of brilliant granules scattered over a darker background (Fig. 1.1). According to Rösch these granules number 3.5 million. The individual grain, with an average diameter of a thousand kilometers, is not a stable configuration but in fact has a short life-span: it appears, shines, and disappears in the course of a few minutes. Often the space between the granules is abnormally wide because some are absent, and a dark, roundish gap, called a pore, forms in the granular network (Fig. 1.2). Pores may last longer than granules, sometimes over an hour.

According to the current interpretation the granules are nothing but the end portions of enormous columns of gas which continually ascend, at a velocity of about a kilometer a second, from the lower zones of the photosphere to higher levels; we see the columns not from the side but from above, in cross section. When a whole group of neighboring granules is absent, we observe in its stead the darker background and a pore appears.

Sometimes, however, vast regions of the photosphere undergo perturbations, the granules disappear, and in their place appear dark zones, having a complicated structure, surrounded by a lighter region, and occasionally crossed by brilliant filaments. These are the famous sunspots (Fig. 1.3),

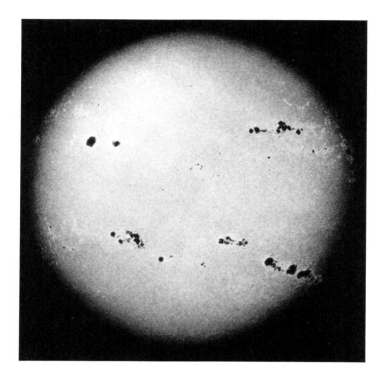

Fig. 1.1   The solar disk as observed through a telescope. The edge appears faint and less bright because of the various degrees of absorption of the overlying gases. Also visible are the granulation and several spots, some isolated, some in groups. (*Mount Wilson and Palomar Observatories*)

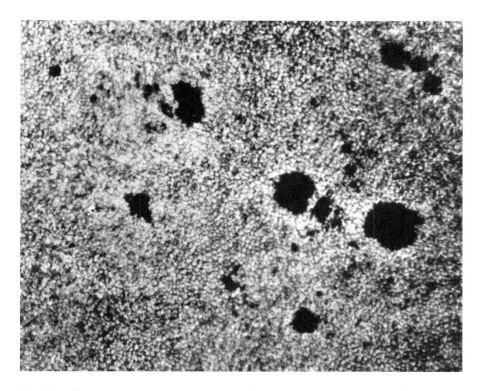

Fig. 1.2   Solar granulation in an exceptional high-resolution photograph taken by Janssen on 5 July 1885; some pores are also visible. (*Observatoire de Meudon*)

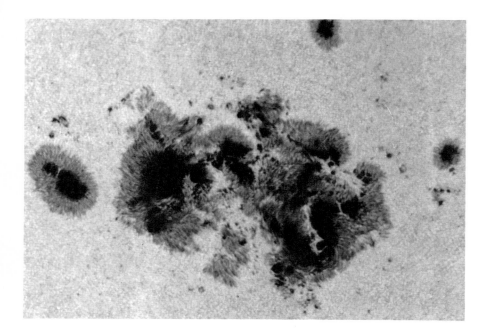

Fig. 1.3   A striking group of extensive and intricate sunspots photographed at Mount Wilson Observatory on 17 May 1951. (*Mount Wilson and Palomar Observatories*)

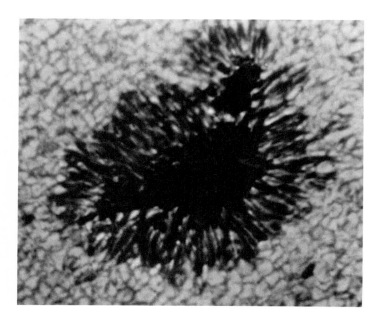

Fig. 1.4   A sunspot photographed at high magnification and resolution: note in particular the ray-like structure of the penumbra. This photograph was taken in the course of the *Stratoscope* observations, with a telescope carried on a balloon above the denser strata of the earth's atmosphere. (*Harvard College Observatory*)

which can be big enough to be visible to the naked eye. The central, darker zone is called the umbra; the outer zone, the penumbra. Observations with telescopes of high resolving power show that the penumbra has a radiating structure; that is, it is composed of a huge number of filaments (having an average diameter of 300 kilometers and a length of about 5,000 kilometers) pointed toward the center of the umbra (Fig. 1.4). These filaments also form and dissolve constantly: each lasts, on the average, no longer than three-quarters of an hour.

Clearly nothing stable can develop from such restless elements. In fact the sunspots themselves undergo continual changes. A spot forms from the union of several pores, grows, multiplies, becomes a group comprising some larger spots surrounded by a multitude of smaller ones; then little by little it closes in and disappears. Once again that part of the solar surface shows no trace of the spot: it appears exactly as it did before the spot formed.

The dimensions of the spots range from a few thousand kilometers to over a hundred thousand; some groups are so large that they could encompass several planets the size of the earth. Durations are extremely variable: they can range from less than a day to more than three months.

In observing a group of spots daily, one notices that they seem to move across the sun's disk (Fig. 1.5). Thus was it discovered, from the first observations at the beginning of the seventeenth century, that the sun rotates on its axis with a period of about a month. By observing spots at different latitudes, it was also found that their rate of rotation was not uniform: a point at the equator on the solar surface makes a complete rotation in 25 days, whereas a point at latitude 40° takes 27. This means that the sun is not a rigid sphere.

But there is more. The sun is not even a sphere with a well-defined boundary, as it appears to the naked eye or in the telescope. It is, rather, an enormous quantity of matter arranged with spherical symmetry and fading outward in the form of increasingly rarefied gases, which, at a great distance from the center, are no longer arranged with spherical symmetry. Thus, the photosphere is not a well-defined surface, like that of the earth, but a gaseous zone of a certain thickness (as the nature of the granules had already revealed), and the gas extends not only beneath this zone but also above it, that is, toward us, even if normally we are unable to see it.

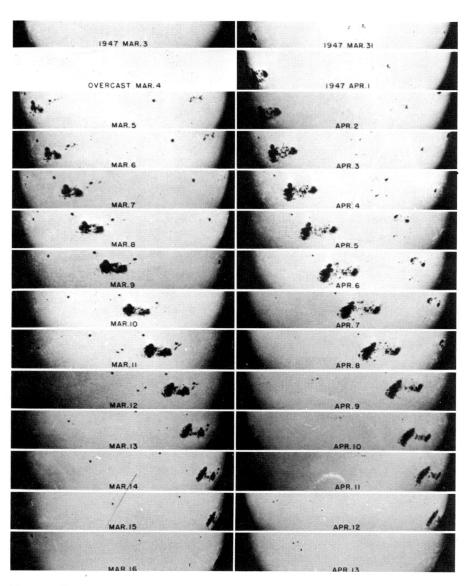

Fig. 1.5   Rotation of the solar disk and evolution of the spots. (*Mount Wilson and Palomar Observatories*)

How is it, then, that the sun appears so well-defined? We shall try to find the answer by means of a simple experiment. Let us take a sheet of cellophane; it appears very transparent. Let us now fold it in two, then in four, and so on, each time doubling the thickness. The combination of all these sheets will become steadily less transparent, and at a certain point we shall see a shiny layer, well defined in appearance, through which we shall see nothing, as if we had inserted a steel blade within the cellophane. Of course there is no steel blade there, but the same effect has been obtained by piling up so many sheets, each of which was not, as it seemed to be, completely transparent.

The same thing happens with the sun. The piling up of so many gaseous layers, which are not completely transparent, produces at a certain point an opaque barrier that the eye cannot penetrate. This is the photosphere. But just as there are sheets of cellophane above the imaginary steel blade, each of which contributes to block a small fraction of the light, but is not alone sufficient to produce the effect of the steel blade, so are there layers of gas, more and more transparent and rarefied, which remain invisible under ordinary observation and do not prevent us from seeing the underlying photosphere. The situation is quite different if the sun is observed with special instruments or at certain moments.

Anyone who has been lucky enough to witness the impressive sight of a total eclipse of the sun will have seen, during the few minutes of totality, a splendid, silvery halo surrounding the black disk of the moon superimposed on the solar disk. This halo, trailing off in splendid plumes, bright as the full moon and extending to a distance of 8 million kilometers from the sun, is the outermost layer of the solar atmosphere: the corona (Fig. 1.6). It consists of an enormous halo of electrons toward the center and a zone of tiny dust particles toward the periphery. Spectroscopic analysis has revealed the presence of various highly ionized elements, which shows that the corona must have a temperature of about a million degrees.

During total solar eclipses one can also see in the lowest part of the corona, adjacent to the black disk of the moon, another layer of the sun's atmosphere that is invisible under normal conditions: the chromosphere (Fig. 1.7). It appears as a thin crimson arc which, when observed through a telescope under high magnification, shows a dense, threadlike structure in constant motion, like grass in a field stirred by the wind. It is a kind of

Fig. 1.6 Two views of the solar corona: *above,* the inner zone photographed on the occasion of the eclipse of 8 June 1918; *below,* the outer corona photographed during the eclipse of 12 November 1966. (*Harvard College Observatory; High Altitude Observatory, Boulder*)

"burning prairie," as Angelo Secchi called it, in which the blades of grass are tongues of incandescent hydrogen about 8,000 kilometers long, stirred by powerful thermal disturbances and by the action of magnetic fields.

Enormous red flames often develop within the chromosphere, rising for thousands of kilometers and then falling hundreds of kilometers away, forming enormous bridges. These flames, essentially composed of hydrogen and helium, are called prominences. Some of them, such as those which form bridges, last longer and evolve more slowly. Others, eruptive in character, burst away from the sun, reaching tremendous heights within a few hours (Fig. 1.8). The highest have been seen to reach over half a million kilometers above the chromosphere.

The thickness of the chromosphere is relatively modest, about 10,000 kilometers, and, as was mentioned earlier, it is composed mainly of hydrogen. But there is more to be said. We have seen that under direct observation it appears red. This is because both chromosphere and prominences radiate light essentially in the so-called Hα line of hydrogen, which is red. Therefore, if we attach to the telescope a red filter, which allows only the light of the Hα line to pass, the rest of the light coming from the sun is excluded and we see only the chromosphere (Fig. 1.9). Thus with this miraculous filter, we are able to see this almost transparent and less luminous layer of the solar atmosphere, which the preponderant light of the underlying photosphere normally prevents us from observing,[2] and to see it not only at the edge, as during total eclipses, but over the whole disk and at any time.

Naturally the same phenomena when projected on the sun's disk will appear differently. For example, a prominence that we saw at the edge as an enormous arc will now be seen from above in the shape of a filament (Fig. 1.10), and from the marked thinness of the filament we shall discover that the giant prominences, tens of thousands of kilometers high and hundreds of thousands of kilometers long, are scarcely six thousand kilometers thick.

The sunspots appear to be limited to the zone of the umbra, but the structure of the overlying chromosphere shows strange configurations, ar-

[2] The construction of filters designed for this purpose has been anything but easy and has progressed notably only in the last decade.

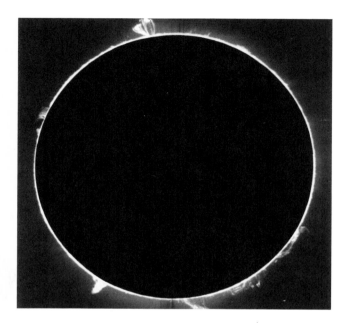

Fig. 1.7 Photograph of the solar chromosphere and of some prominences, about 100,000 km high, taken 9 December 1929. Because of the low enlargement in this picture, it is not possible to distinguish the threadlike structure of the "burning prairie." (*Mount Wilson and Palomar observatories*)

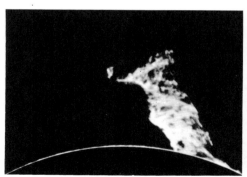

Fig. 1.8 The evolution of a prominence in the course of 29 min. It is seen falling back on the sun's surface. This was photographed in the light of the Hα line of hydrogen. (*Sacramento Peak Observatory*)

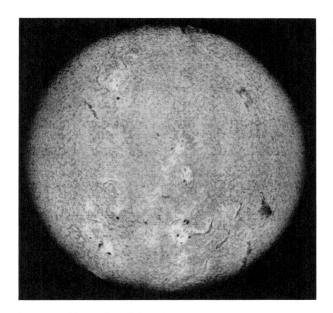

Fig. 1.9   The solar disk as it appears when photographed in the light of the Hα line of hydrogen. The dark filaments are prominences projected on the luminous disk, and their thinness shows that these enormous flames are very narrow in cross-section. (*Mount Wilson and Palomar Observatories*)

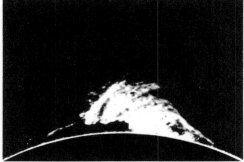

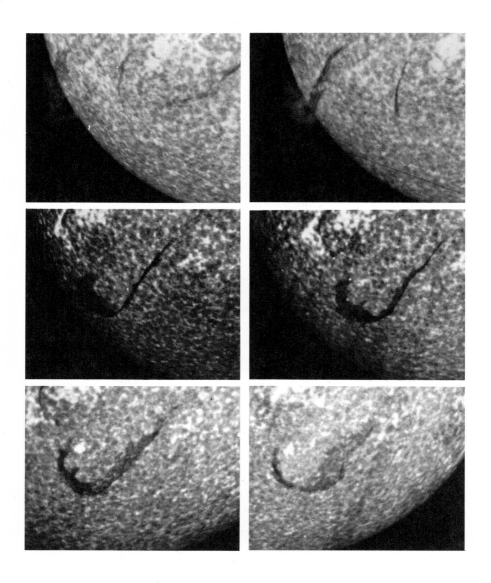

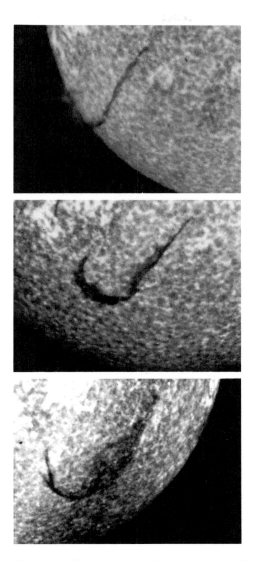

Fig. 1.10   The evolution of a prominence whose appearance gradually becomes that of a filament. On account of the solar rotation, the prominence is shown projected on the disk. Note the evolution of the prominence-filament as time passes. The sequence reads from left to right and from top to bottom. (*Observatoire de Meudon*)

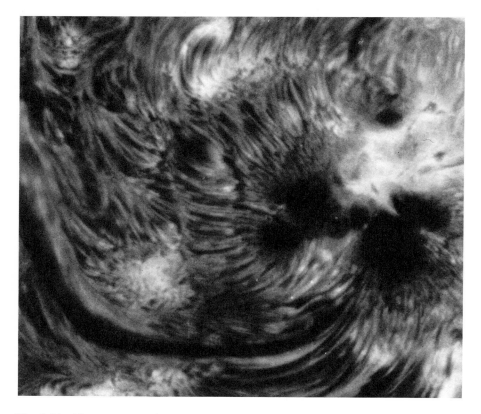

Fig. 1.11 Vortex structures of the solar chromosphere in a storm; they suggest the presence of magnetic fields, which were discovered and measured by means of spectroscopic observations. The photograph was taken in the red light of the Hα line of hydrogen. (*Big Bear Solar Observatory*)

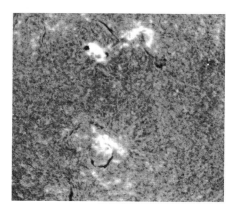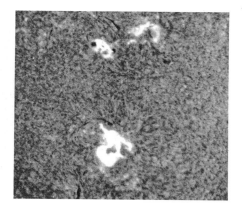

Fig. 1.12 Solar flare photographed at Catania Observatory on 9 June 1968. *Left:* Taken at 08:25, this photograph shows the disturbed region where the flare will appear in a quarter of an hour. *Right:* Taken at 08:48, this photograph shows the developed flare. The phenomenon vanished at 11:20. (*G. Godoli*)

ranged like the lines of force in a magnetic field (Fig. 1.11). In fact it has been discovered that sunspots are sites of intense magnetic fields, which range from 100 gauss for some pores to 4,000 gauss for certain large groups.

In a disturbed zone of the sun, the site of spots or prominences, there appears another phenomenon, which is sudden and of brief duration: the flare (Fig. 1.12).

Flares are sudden flashes of light, sometimes so intense as to exceed the brightness of the photosphere itself. In these exceptional cases they can be observed even through a normal telescope, that is, without recourse to the filter technique used to observe the chromosphere. The phenomenon develops with impressive rapidity: it can start and grow in a very brief time—a flash (like a bolt of lightning, perhaps) can extend in a few minutes over a surface ranging from fifty million to a thousand million square kilometers. After a few hours, or even after a few ten-minute periods, it will have disappeared entirely. Often, however, the first flash is followed by others in rapid succession. In some fully developed groups of spots up to some fifty flashes have been observed.

Solar flares have been studied regularly for some twenty-five years. It is only recently, however, that a discovery of great practical importance—above all, in relation to human voyages in space—has been made. It was found that certain flashes from the sun produce high-energy protons and cosmic rays with relativistic energy (Fig. 1.13) that flood the earth a few minutes or a few hours later.

Thus the sun strikes the earth not only with its beneficent rays, bringing light and heat, but also with deadly radiations that penetrate interplanetary space, interacting with the matter scattered there. For those dwelling on the surface of the earth there is no danger, since the atmosphere constitutes a protective shield, but for astronauts traveling in space or on the moon, who are struck by the full primary radiation emitted directly by the flare, the effect may be fatal. One will recall that there was some anxiety concerning the astronauts of *Apollo 12,* precisely because during the second landing on the moon some proton flares were observed on the sun.

But let us leave to biologists the study of this aspect of the subject and turn to the sun to inquire, first of all, "What are flares?" Even if we cannot yet give a definite answer, we can put forward an interpretation that may be, at least qualitatively, the correct one. We have seen that the zones where flares appear are the sites of intense magnetic fields. We know that when a wire carrying an electric current is suddenly torn, the abrupt break in the magnetic field associated with it causes a very brief, blinding luminous discharge. It may be that something of the sort takes place in the magnetic fields associated with disturbed zones on the sun.

On the other hand, the existence of the spots themselves, of the prominences, and of all the phenomena associated with solar activity has not yet met with a complete and satisfactory explanation.

As we have seen, when we observe the sun with even a modest telescope, a certain number of spots, isolated or in groups, are visible. If we continue to observe the sun over months and years, we notice that the average number of these spots does not always remain the same, that there are long periods when no spots are seen, and other times when the sun shows many spots, some even big enough to be seen with the naked eye. The phenomenon is cyclical in character and lasts eleven years (Fig. 1.14). In recent times, for example, 1947, 1958, and 1969 were years of maximum activity, and 1944, 1954, and 1964 were years of minimum activity. The same cycle holds, albeit in different ways, for prominences, flares, and the form of the corona. The periods of maximum activity are called periods of the active sun; those of minimum activity, periods of the quiet sun. During these cycles other specific phenomena are observed with regard to the positions of the spots and the associated magnetic fields. Without going into detail, it is enough to add that these phenomena enable us to determine that the true cycle is double, that is, lasting twenty-two years, and that it is certainly more complex than telescopic observation of the sunspots alone would indicate.

IN THE SUN'S INTERIOR

All that we have seen up to now is what happens, in the short and long run, in the zones from which the enormous amount of energy whose benefits we enjoy is emitted. The sun has been pouring energy into space for thousands of millions of years. We have not yet explained, however, where or how this energy is produced. To clear up the mystery, we shall have to descend into the sun's interior.

Let us start from the photosphere, the layer that emits most of the light and heat that escape from the sun to outer space. This layer, essentially composed of hydrogen at a temperature of 6,000°K, must receive heat from the underlying region, which must be not only hotter but also denser, since it is compressed by the outer layers. Repeating this reasoning for ever deeper layers, we come to view the sun as a gaseous sphere that grows denser and hotter the farther we advance from the outside toward the interior regions.

These ideas, viewed quantitively and in conjunction with the fundamental laws of physics (which are assumed to be valid even in the sun), led some theoretical astrophysicists, at the end of the last century and in the first half of the present one, to construct schemes and models that provided a reasonable interpretation of the interior regions of the sun.

By basing hypotheses only on the *distribution* of the energy sources and leaving aside for the moment the question of their origin, various models were derived. One of the most reasonable seemed to be that the energy originates only in a limited zone in the center of the sun and then moves from one to the other of the overlying layers, until it escapes outward. In this way important values for the temperature and the pressure were found.

According to one of the most widely accepted models, the temperature in the central zone of the sun would be twenty-five million degrees; the pressure, two hundred thousand million atmospheres; and matter would have a density 110 times that of water. By changing the model, one obtains different numerical results which nevertheless have the same order of magnitude. In other words the temperature could be twenty or fifteen million degrees, but not twenty thousand.

Thus we know for certain what the physical conditions of the interior of the sun must be, but we still have not explained the origin of the energy that produces such a high temperature or, most important, how the sun has continued to expend this energy for thousands of millions of years.

If the sun were made of coal, it would be reduced to ashes in a thousand years. Obviously, then, the energy cannot be produced by any of the processes of combustion known to us. Toward the end of the last century the physicists H. L. von Helmholtz and W. T. Kelvin conceived another mechanism, based on the transformation of mechanical energy into heat. It is a well-known fact that this process is always possible and that we apply it

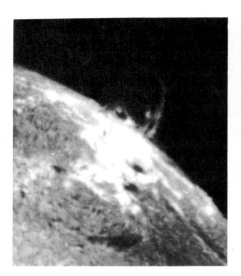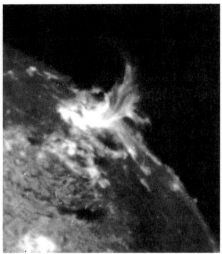

Fig. 1.13   The explosion in sequence of matter associated with a flare. The photos were taken in the red light of the Hα line of hydrogen on 15 May 1957, at 15:51, 15:56, and 16:10, respectively. The jets are easily seen since the phenomenon was observed in profile toward the edge of the sun. (*Fraunhofer Institut*)

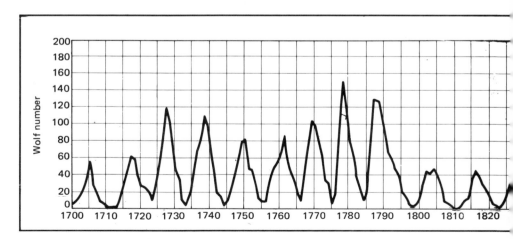

Fig. 1.14   Periodic variations of solar activity in the period 1700–1970. The Wolf number expresses the solar activity as a function of groups and single spots.

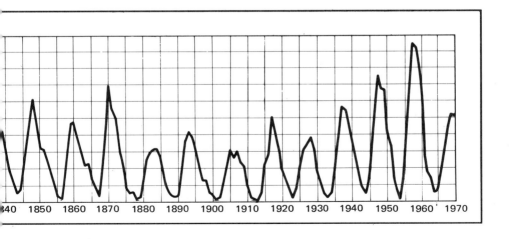

quite frequently—for example, when we rub our hands to warm them or when we burn out the brakes of our car by constant braking.

Let us imagine the following experiment: Suppose we drop a stone into a pond. Being subject to the force of gravity, the stone will fall with a certain velocity and hence has a certain energy. (We might perceive this fact more readily if, instead of falling into the pool, the stone were to drop on one of our feet.) Well, we may ask, what happens to that energy when the stone reaches the bottom of the pond and stops? The answer is simple: It is absorbed by the particles of water in the pond, which then move with a certain velocity. If we continue to throw stones into the pond, the velocity of the water particles will increase steadily, and this will increase the temperature of the water since, as a physicist would tell you, the temperature of a fluid is determined by the velocity of its tiny, invisible constituent particles.

This process is clearly not the most practical method for heating a quantity of water, but according to Helmholtz and Kelvin it could work well for the sun. In other words they assumed that entire outer layers of the sun fall inward. The gravitational energy is then transformed into heat, and the temperature of the underlying layers is raised. The mechanism is still that of the stone falling into the pool, but the mass in question is far greater, and as a result the energy that can be derived is enormous.[3]

Thus, by progressively contracting, the sun could produce a quantity of energy ten thousand times greater than that provided by chemical combustion. Unfortunately, even in this manner the sun could not survive more than ten million years, whereas the existence of fossils proves that the sun has been heating our planet for hundreds of millions of years as it does today. Thus was this theory shown to be inadequate to explain the observed facts.

The beginning of the correct approach was found for the first time in 1927 by two young physicists of the celebrated Göttingen school: R.d'E. Atkinson and G. F. Houtermans. It was a new approach, a bold and stimulating one, that led the two physicists to explain the origin of solar heat but would subsequently lead all mankind to the most fearful crossroad in its entire history, because it was also the most direct approach to the H-bomb.

Thus we approach the culmination of modern natural philosophy. The

[3] The energy that a moving body possesses (and can therefore surrender) depends not only on its velocity but also on its mass. Anyone not convinced of this fact should imagine being hit by a fly or by a truck, both moving at the same speed.

atom and indeed the still smaller particles that compose it are about to enter and read us the riddle of the universe.

In the past some scientists launched out toward the infinite; others descended to the infinitesimal. They had discovered the two abysses between which man is suspended while participating in both. Now the infinitely small escapes from our laboratories to reappear inside the stars as the creator of the immense world that confronts us when we observe the heavens. The particle explains the universe to us, for the universe is made up of particles, which join together to form the atom; the stars, where the elements are born; interstellar space, where the elements unite to form molecular compounds; the planets, where molecules can combine in such a way as to form life. And this life, for us the greatest and most mysterious of phenomena, rises from the unicellular being to the plant, to the animal, to Man, who attempts through his intelligence to contemplate and understand the universe; and, higher still, to creative genius. That genius is one of the greatest gifts of God to Man, the interpreter of His work.

There is no grander synthesis than this; and its starting point is there, inside a star, inside an atom, which we must now explore if we wish to continue our journey toward the infinite.

THE CRUCIBLE OF THE ELEMENTS

According to the ancients, the atom, the smallest part of an element to retain its properties, was indivisible. Modern physics has taught us, however, that even the atom is composed of smaller particles, differing in mass and electric charge, which form, according to the number combined, the various chemical elements. Matter is composed of three fundamental particles: electrons, protons, and neutrons.[4]

The electron has a negligible mass and an appreciable negative electric charge. The proton has a considerable mass (1,836 times that of the electron) and an electric charge equal to that of the electron but positive. The neutron has the same mass as the proton but has no electric charge, that is, it is electrically neutral.

The nucleus of every atom is composed of a certain number of protons

[4] Many other elementary particles exist, but since they do not affect the essence of my argument, they will not be listed here.

and neutrons, and it is thus electrically positive. Around the nucleus, at a certain number of levels, comparable with the orbits of a planetary system, there rotate a certain number of electrons, equal in number to the protons, which balance the electrical charge of the nucleus. Thus the normal atom of an element is electrically neutral. The number of protons present in the nucleus determines the element: hydrogen, which is the simplest, is characterized by the presence of one proton; helium has two; lithium three; and so on, up to uranium, the most complex natural element, which has ninety-two. Normally the number of protons present in the nucleus is matched, externally, by an equal number of electrons. Thus the normal hydrogen atom has one electron; the helium atom, two; and the uranium atom, ninety-two. Sometimes, however, one or more electrons may be missing. Then the atom is no longer electrically neutral but is positively charged to some degree, depending on the number of electrons lost. In such a case the atom is said to be ionized once, twice, or several times, depending on whether it has lost one, two, or more electrons. Of course, even in this case, the atom still belongs to the same element; for example, an iron atom that has lost two electrons is still an iron atom.

The number of protons present in the nucleus partly determines the mass of the atom since, as we have seen, the mass of the electrons is negligible. The remaining contribution to the mass of the atom is determined by the number of neutrons, which, having no charge, nevertheless have masses equal to those of the protons. The number of neutrons in the nucleus of an atom of a given element is not always the same. It is possible, therefore, to have variant atoms of an element with different masses. These variant atoms are called isotopes. Thus the normal hydrogen atom, as we saw, is composed of a nucleus with a single proton and an electron. One isotope of hydrogen has a nucleus composed of a proton and a neutron; another much rarer isotope has a nucleus composed of a proton and two neutrons. The first of these isotopes is called heavy hydrogen; the second, tritium.

At this point we cannot help noticing that we have at hand the key to realizing the dream of medieval alchemists: that of transforming one element into another. In theory, at least, this process is simple. If an element is determined by the number of protons present in its nucleus, we need only add or subtract protons in order to transform it into another element. For example, the nucleus of a lithium atom is composed of three protons and four neutrons; thus if we add a hydrogen nucleus, that is, a proton, we can have a nucleus composed of four protons and four neutrons—or better yet, two helium nuclei, each of which is composed of two protons and two neutrons (Fig. 1.15).

Realizing this transformation is, however, extremely difficult—at least as long as matter appears under the physical conditions in which it is normally present on our planet. But if, by means of special devices, we do succeed in realizing this transformation, we discover a very strange phenomenon. If we take the necessary number of nuclei of a light element (for example, hydrogen or helium) and manipulate these nuclei so as to construct the nucleus of a heavier element, we find that the sum of the masses of the nuclei used is not equal to that of the nucleus obtained, which turns out instead to be somewhat lighter, and the heavier the nucleus obtained, the less tends to be this difference in mass. These facts hold for all the elements lighter than iron. For those elements heavier than iron, however, the opposite happens: that is, the nucleus of a heavier element formed by the combination of two lighter ones now has a greater mass than the sum of the masses of the two nuclei from which it is formed.

This phenomenon contains the key to nuclear bombs and the solution to the mystery of the origin of the heat inside the sun. More precisely, the key is found not only in this fact but also, essentially, in one of the fundamental formulas of relativity theory.

It is a very simple formula, and contained in it is the whole destiny of mankind, whether prosperity or self-destruction:

$$E = mc^2$$

The symbols denote energy (E), mass (m), and the velocity of light (c). This formula expresses the equivalence of mass and energy and means that a body of a given mass contains in itself an amount of energy equal to the value of the mass multiplied by the square of the velocity of light. Given that the velocity of light is represented by a very large number, it is clear that even a very small mass is equivalent to an enormous amount of energy. For example, the energy, in electron volts, that a mass of one gram can produce is expressed by the number 56 followed by thirty-one zeros. This explains what happens to the small fraction of the mass that disappears when we form an atom from the union of two lighter elements or split an

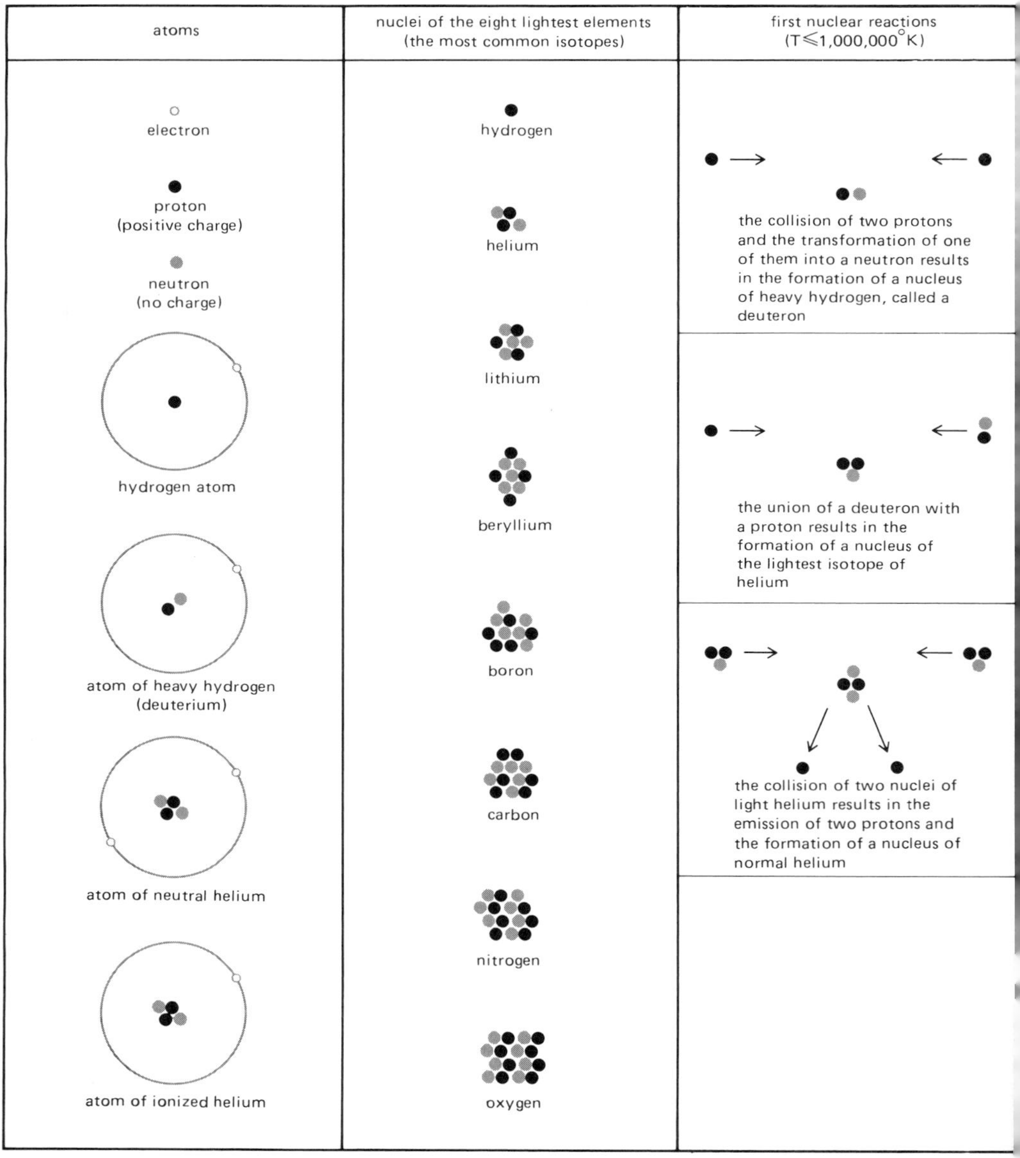

atoms	nuclei of the eight lightest elements (the most common isotopes)	first nuclear reactions (T ≤ 1,000,000°K)
electron	hydrogen	
proton (positive charge)	helium	the collision of two protons and the transformation of one of them into a neutron results in the formation of a nucleus of heavy hydrogen, called a deuteron
neutron (no charge)	lithium	
hydrogen atom	beryllium	the union of a deuteron with a proton results in the formation of a nucleus of the lightest isotope of helium
atom of heavy hydrogen (deuterium)	boron	the collision of two nuclei of light helium results in the emission of two protons and the formation of a nucleus of normal helium
atom of neutral helium	carbon	
atom of ionized helium	nitrogen	
	oxygen	

Fig. 1.15 *Left to right:* Lighter atoms, lighter nuclei, first nuclear reactions, reactions of lithium, beryllium and boron, carbon-nitrogen cycle. All these processes lead, in fact, to the formation of helium nuclei. In particular, those relating to lithium, beryllium, and boron irreversibly consume these elements.

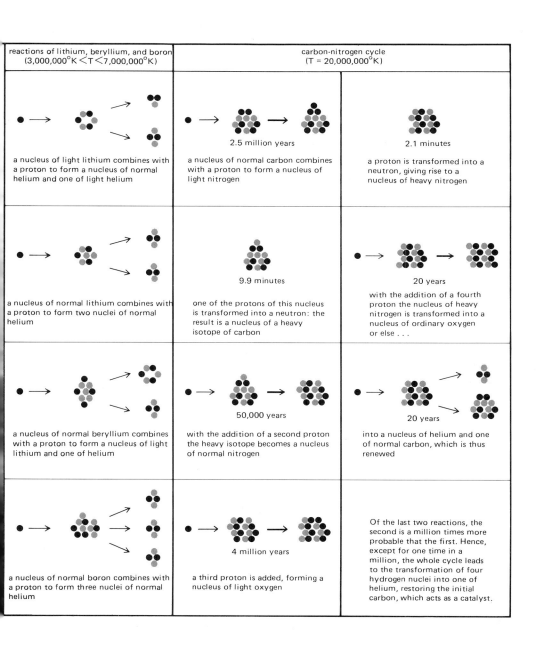

reactions of lithium, beryllium, and boron ($3{,}000{,}000°K < T < 7{,}000{,}000°K$)	carbon-nitrogen cycle ($T = 20{,}000{,}000°K$)	
a nucleus of light lithium combines with a proton to form a nucleus of normal helium and one of light helium	2.5 million years — a nucleus of normal carbon combines with a proton to form a nucleus of light nitrogen	2.1 minutes — a proton is transformed into a neutron, giving rise to a nucleus of heavy nitrogen
a nucleus of normal lithium combines with a proton to form two nuclei of normal helium	9.9 minutes — one of the protons of this nucleus is transformed into a neutron: the result is a nucleus of a heavy isotope of carbon	20 years — with the addition of a fourth proton the nucleus of heavy nitrogen is transformed into a nucleus of ordinary oxygen or else . . .
a nucleus of normal beryllium combines with a proton to form a nucleus of light lithium and one of helium	50,000 years — with the addition of a second proton the heavy isotope becomes a nucleus of normal nitrogen	20 years — into a nucleus of helium and one of normal carbon, which is thus renewed
a nucleus of normal boron combines with a proton to form three nuclei of normal helium	4 million years — a third proton is added, forming a nucleus of light oxygen	Of the last two reactions, the second is a million times more probable that the first. Hence, except for one time in a million, the whole cycle leads to the transformation of four hydrogen nuclei into one of helium, restoring the initial carbon, which acts as a catalyst.

element heavier than iron to make two lighter ones: the missing mass is transformed into energy.

Thus if we succeed in splitting the atom of a heavy element—for example, uranium—to form two lighter ones (in which, we repeat, the sum of the masses will be less than that of the uranium), the missing mass escapes by transforming itself into energy. This is precisely what happens in the atom bomb.

On the other hand, if we succeed in joining lighter nuclei to form heavier ones (whose mass will be less than the sum of those of the original nuclei), the missing mass is transformed, again, into energy. This is what happens in the hydrogen bomb.

The breakup, or fission, of the uranium atom can be accomplished more simply than the fusion of light nuclei because the latter takes place only at extremely high temperatures that do not exist on the surface of the earth but which, as we have seen, are entirely normal in the sun's interior.[5] Astrophysicists have discovered various thermonuclear reactions that can exist in the sun's interior, each of which requires a certain optimum temperature. At a temperature of a million degrees it is possible to obtain heavy hydrogen from the union of two protons. At temperatures between three and seven million degrees, the fusions of protons of lithium, beryllium, and boron take place in the order of increasing temperature. Finally, at a temperature of about twenty million degrees, the so-called carbon-nitrogen cycle develops, in which four hydrogen nuclei are transformed into one helium nucleus, after passing through the formation and breakup of carbon and nitrogen, which are nonetheless present in their original amounts at the end of each cycle. Thus have we found the key to the origin of energy within the sun, and we have discovered not only that, but also how all the elements except one are formed. This exception is the simplest, lightest, and most abundant element in the universe: hydrogen.

Heavy hydrogen is formed, in fact, from the proton + proton reaction. The fusion of heavy hydrogen and normal hydrogen produces helium. In this

---

[5]Human ingenuity has managed, however, to bridge this gap. Using an atom bomb (that is, a uranium bomb) as a detonator, one can in fact reach a temperature of the order of some millions of degrees, which is sufficient for the processes of fusion of light elements to take place, thus making possible the hydrogen bomb with its capability of destroying mankind much more rapidly and efficiently.

manner heavier and heavier elements are created, through increasingly complex reactions that require increasingly higher temperatures. These reactions in turn develop an enormous amount of energy, which is partly dissipated but also partly contributes to the rise in temperature of the reacting matter, speeding the reactions and rendering possible others that demand higher temperatures. This process goes on as far as iron. Starting with this element, however, it is not possible to fuse nuclei by extracting energy but, on the contrary, fusion can be obtained only by supplying energy. It is believed then that, beginning with the elements somewhat lighter than iron up to the heaviest ones known, the mode of formation is different. The places where the heaviest elements are produced seem today to be well identified, and we shall find them later on.

In this whole argument there is still one point that is not very clear. We have seen that the processes of fusion can forge the elements starting with hydrogen and at the same time produce energy. But we have also seen that for these reactions to take place temperatures of at least a million degrees are needed. How then is this very high temperature initially produced, which makes possible the first thermonuclear reactions?

The answer has already been given by Helmholtz and Kelvin: by means of gravitational contraction. This theory, which was not sufficient to explain the production of solar energy for thousands of millions of years, can instead be very well applied to the early phases in the life of the sun, when it was an enormous and amorphous cloud of hydrogen. At that epoch the cloud contracted under its own weight and, falling in towards the center, developed energy, part of which helped to increase the temperature. The contraction continued until the temperature was high enough to make possible the priming of the first nuclear reactions. Then the energy produced in the sun's interior increased to the point where it balanced the fall of matter inward, the gravitational contraction stopped, and energy continued to be produced with the fusion processes, which continue today.

WHAT THE SUN IS

Now that we have descended into its interior, into the immense furnace where heat is produced and the elements are forged, we can finally form a complete idea of what the sun is (Fig. 1.16).

It is a fiery sphere where, at very high temperatures and at pressures that we can express only in numbers that transcend our imagination, an enormous quantity of energy is developed. A large part of this energy reaches the surface, passing through incandescent layers (albeit at temperatures a great deal lower than those in the center), where the gases, subjected to a multiplicity of forces, form immense hurricanes that produce brilliant explosions and gigantic eruptions. Above the flames of the chromosphere extends the silvery corona, where this monstrous body, this colossal, constantly exploding thermonuclear bomb fades away into space in a delicate cloud of pearly light. But perhaps the corona does not end at the point where, as seen by the eye and photographed during total eclipses, it seems to end. At the time of the equinoxes, one can see, soon after sunset in the western sky or shortly before dawn in the eastern sky, a luminous cone as bright as the Milky Way, which, if extended, would reach the sun. This is the so-called zodiacal light, a luminescence due to the diffusion of the sun's light by dust scattered in the solar system, far beyond the corona, of which it is, in a certain sense, the continuation. This dust goes beyond the earth's orbit, which indicates that the outermost layers of the sun reach that far. Therefore, we can define the sun more concisely as a colossal furnace that radiates energy into space in all directions and extends materially for millions of kilometers, fading away in increasingly rarefied veils.

And there are hundreds, millions, thousands of millions of stars like the sun. We need only look up at the sky on a clear night to see a limitless host of them. Because every star is a sun.

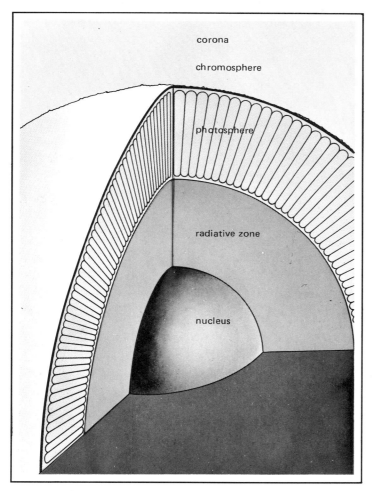

Fig. 1.16   Section of the sun. From the outside: The corona, very extensive and without sharp boundaries; the chromosphere, 10,000 km thick; the photosphere, in which the convective processes that produce the granulation are shown; the radiative zone, so called because the heat is transmitted by radiation; the nucleus, in which the thermonuclear reactions of fusion, which are the source of the sun's energy, take place.

THE PLANETS NEAR THE SUN

The sun, the first stop in our journey, is not only an enormous machine in which the elements are manufactured and energy is produced but also the center of attraction for a vast system of heavenly bodies, which revolve around it at various distances and receive from it light and heat. One of these bodies is the earth, the world that we inhabit.

The earth is one of nine planets, all different in size, that orbit the sun (Figs. 2.1 and 2.2). In addition to these larger bodies, the solar domain also comprises hundreds of minor planets, numerous comets, hosts of meteorites, and dust clouds. That part of space in which all these bodies move is the part we shall explore first, beginning with the major planets, which we shall visit one by one. We start with the planet nearest the sun and move outward, to the limits of the solar system, where we find Pluto, the most distant planet yet known.

MERCURY

The planet nearest the sun is Mercury, a sphere with a diameter of 4,880 kilometers—larger than the moon but much smaller than the earth. Its mass is barely 5.4 percent that of the earth; its density is 5.5 times that of water. It revolves around the sun at an average distance of 57.9 million kilometers, completing one revolution every 88 days.

Mercury rotates on its own axis, but very slowly. Until a few years ago the commonly accepted value was 88 days, as found by G. V. Schiaparelli in the late nineteenth century and confirmed by subsequent observations. If that were the case, its periods of revolution and rotation being equal, Mercury would always show the sun the same hemisphere, as the moon does to the earth. In 1965, however, scientists at Arecibo found Mercury's rotation period to be 59 days. They did this by sending radio impulses toward Mercury and studying the behavior of the waves it reflected. Today the accepted period of rotation for Mercury is 58.65 days, proposed by G. Colombo and recently confirmed by other radar observations and by *Mariner 10*. With this period the planet makes exactly three rotations about its axis during two revolutions around the sun.

In accordance with this new result, Mercury does not always present the same hemisphere to the sun; however, given the length of the period of

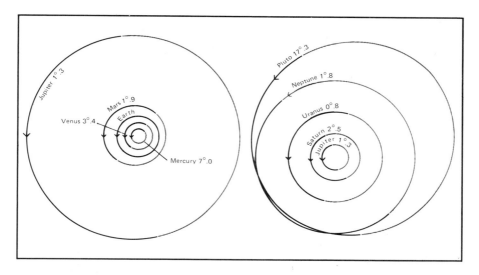

Fig. 2.1 *Left,* inner region, and *right,* the outer region of the solar system. The planets all move in the same direction, in orbits that are almost circular and are practically in the same plane, except for that of Pluto, which is inclined at 17°.3 to the plane of the Earth's orbit (the ecliptic). The inclination of the other planets with respect to the ecliptic is indicated beside the name of each; the part of the trajectory that is above the ecliptic is shown in a darker shade and that below in a lighter one. The orbits of the asteroids (not shown here) are much more eccentric than those of the planets and lie between the orbits of Mars and Jupiter (see fig. 2.15). The inner region of the solar system is shown here on a scale 7.5 times greater than that of the outer region.

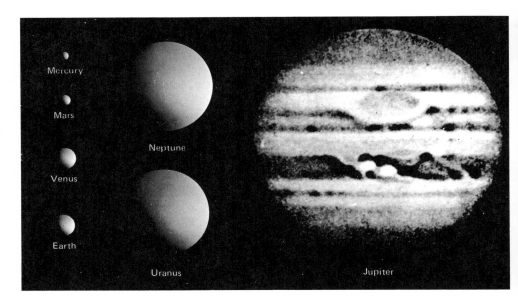

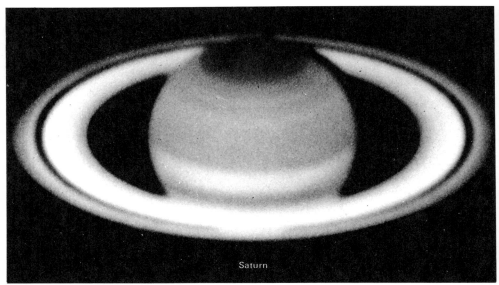

Fig. 2.2   Comparison of the planets of the solar system shown to scale. Pluto is not shown, because its dimensions are uncertain.

rotation, the alternation of day and night is very slow, and the difference in temperature between the two hemispheres must be very great. In fact, according to the most recent observations, the temperature of the illuminated hemisphere at about midday is + 350°C, while that of the dark side is −170°C.

Obviously, the primary reason the temperature is so high on the illuminated hemisphere is that the planet is so close to the sun; moreover, the enormous drop in temperature between the bright and dark sides suggests another cause—the absence of an atmosphere.

The lack of a gaseous envelope was proved by the *Mariner 10* probe, which in 1974 and 1975, in three successive rendezvous with the planet, photographed a large part of Mercury's surface. In the photographs transmitted to earth, the ground appears similar to that of the moon—riddled with craters, large and small (Fig. 2.3). Careful examination, however, has shown certain differences between the surface of Mercury and that of the moon: the number of craters scattered over the surface is higher. Furthermore, they often appear in groups or in sequences of aligned craters. Escarpments have been found that seem to have been formed during a long period in which the planet's crust shrank, perhaps as a result of the cooling and contraction of a large iron core which is thought to constitute the planet's internal structure.

Neither the oldest craters (which are thought to go back 3,000 to 4,000 million years) nor, a fortiori, the more recent ones show any trace of surface erosion. This fact proves that since at least the very remote epoch when the first craters were formed Mercury has not had even a tenuous atmosphere.

Mariner 10 also discovered the existence of a magnetic field, very weak (.0037–.0070 gauss) but sufficient to explain the presence of a very thin layer of helium gas surrounding the planet. This tenuous veil consists of helium emitted by the sun and captured by the magnetic field. Clearly this layer of helium could not be called an atmosphere, for it is incapable of moderating, even slightly, the merciless rays of the sun.

The temperature distribution on Mercury's surface is rather peculiar. As one passes from day to night, there is an enormous jump, all the greater the nearer one is to the equator. But even along the equator there are considerable differences of temperature, depending on the longitude; the combina-

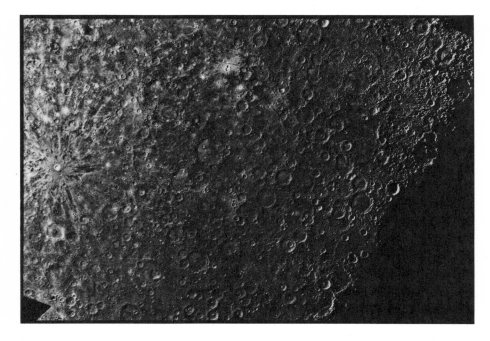

Fig. 2.3 A general view of the surface of Mercury, photographed from *Mariner 10* and obtained by combining various photographs and subjecting the final image to a process of smoothing, so as to eliminate the distortions in perspective due to the curvature of the planet's surface. Note the marked resemblance to the lunar terrain.

tion of the motion of revolution with that of rotation produces this unusual effect in the following manner. At longitudes 90° and 270° the sun rises, reaches its zenith and sets regularly (albeit very slowly, because of the length of the Mercurian day). At these two longitudes "midday" occurs at aphelion, that is, at that point in the orbit where the planet is farthest from the sun and, accordingly, moves most slowly. At longitudes 0° and 180°, on the other hand, midday occurs at perihelion, where the planet is nearest to the sun and the orbital motion is most rapid; at these points, the orbital angular velocity is greater than the angular velocity of rotation.

Because of this last fact, if we were at longitudes 0° or 180°, we should witness a strange phenomenon. As midday approaches, the sun slows down in its course across the sky and actually reverses direction; after retrogressing a short distance, it appears to stop once more; then it proceeds to move forward, at first slowly then with increasing speed until it sets.

This "slowing-down" of the sun, at the hottest time of the day and during the very days when Mercury is closest to it, makes the regions near longitudes 0° and 180° particularly torrid. On these days the temperature rises to 430°C, and the ground must be so red hot that even below the surface the temperature must be above 0°C. In the polar regions, on the other hand, the subsurface temperature is believed to be much below the freezing point of water. Thus it is possible that there may be water in the subsoil of these regions even on torrid Mercury. Moreover, small craters, with sides so steep that the bottoms are permanently sheltered from the sun, have been discovered near the poles. If in some remote epoch volatile substances such as carbon dioxide or water escaped from Mercury's interior, these regions could have trapped them and preserved them in the form of ice to this day.

Apart from these details, the conditions on Mercury are very like those on the moon: it is an arid world without life. The surface is scorched by an enormous sun, whose apparent diameter varies, according to its distance, from 67' to 104' of arc; that is, it appears from 2 to 3½ times as large as it does when viewed from the earth.

Let us, therefore, without regret, leave this world, which is so inhospitable to the body and uninspiring to the mind. Moving farther from the sun, let us approach the second planet, Venus.

VENUS

At a first glance Venus seems to be a heavenly body very similar to the earth. It has a diameter of 12,235 kilometers, and a mass nearly equal to that of the earth (81.5%). Its average density relative to water is 5.2, only slightly lower than that of our planet. It moves around the sun at a distance of 108 million kilometers, making a complete revolution every 224.7 terrestrial days. Some of the early observers also thought that Venus resembled the earth in the length of its period of rotation about its polar axis, which was thought to be about twenty-four hours. On this very point, however, the first substantial disagreements arose, for other astronomers found much longer periods of rotation, in complete disagreement with the values found earlier.

The cause of this controversy resides essentially in the fact that Venus, observed through a telescope, shows no precise reference points—unlike the moon, which has no atmosphere, or Mars, which has a very thin one. Because of its position relative to the earth and sun, the disk of Venus, when seen from our planet, shows phases similar to those of the moon. It appears very white, and only at certain moments can a particularly sharp eye discern in it some dark shadings lacking definite outlines. For some time astronomers have discussed at length whether these spots are stable and, above all, whether they can be considered details of the solid crust beneath a thick atmospheric cover. Today, by means of spectroscopic observations, radar contacts with the planet's surface, and the explorations made by space probes, we are certain that, like the earth, Venus has a solid crust but is, in addition, surrounded by a heavy atmosphere.

Through radar contacts made in recent years it has been found, first of all, that the diameter of the planet is 12,106 kilometers—less than the figure determined from previous optical observations. Obviously the latter figure includes the atmosphere, which reflects the sun's light before it can reach the surface; the figure obtained by radar refers to the ground itself, which the radar waves must strike in order to be reflected to the earth. Since the difference between the two diameters is 129 kilometers, the thickness of the atmosphere must be about 65 kilometers. Radar contacts with the ground have also led to the discovery that Venus revolves backward around its own axis in 243 days. This result is in complete disagreement with the period of barely 4 days obtained from the atmosphere and confirmed by observations made in February 1974 by the *Mariner 10* probe, which photographed the planet from a distance of 5,800 kilometers.

Only one explanation would reconcile the two results: the presence of very strong winds, which make the upper atmosphere (or at least the upper portion that has been photographed) move relative to the lower atmosphere or, indeed, to the ground. From the difference between the two rotations it would follow that the winds have a velocity of about 400 km/h. In fact the Soviet probe *Venera 8,* which landed on Venus on 22 June 1972, measured a velocity of over 360 km/h at a height of 48 km. As the craft descended, however, the wind steadily lessened: between 40 and 20 km it blew constantly at velocities of around 120 km/h; between 14 and 10 km the speed fell rapidly, at the rate of about 18 km/h for every km of altitude; and between ground level and a height of 10 km a wind speed of barely 2 km/h was registered. Thus the high speeds prevail only in the upper strata of the atmosphere, and they are almost certainly the cause of the strange cloud formations photographed by *Mariner 10* (Fig. 2.4).

The *Mariner 10* probe was the first to show us Venus as it appears up close, but it has not been the only one to explore Venus. Between 1961 and 1975 fourteen probes (ten Russian and four American) were sent to Venus. Important discoveries were made with regard to the atmosphere and even the ground: several Russian probes succeeded in landing. The probes confirmed that the atmosphere is much thicker and denser than the earth's and prevents us from seeing the ground from afar; they have also made it possible to analyze the atmosphere and to determine its physical state.

The predominant gas is carbon dioxide. That much had already been determined from spectroscopic observations made in 1932 and was confirmed by the first *Venera* probes, which established that carbon dioxide constitutes 97% of the atmosphere and the remaining 3% is divided among three other components, the main one being carbon monoxide. There is very little water vapor, only 0.1–0.2%, and it is thought to be concentrated in the lowest portion of the atmosphere. Though at first there appeared to be no oxygen, *Mariner 10* has shown it to be present, together with hydrogen, but only in the outermost strata.

For us such an atmosphere would be unbreathable. Moreover, the physical conditions of the atmosphere are such that we could not live there, even

Fig. 2.4   The planet Venus photographed in ultraviolet light from *Mariner 10*. The strange distribution of the upper clouds is clearly visible. The dark portions correspond not to the ground but to the lower regions of the atmosphere, which are no longer transparent.

if enclosed in an aerated diving chamber. At the top of the highest clouds—that is, some 80 kilometers above the ground—the temperature is around −20°C and the pressure is hardly .005 atmosphere. But as one descends, temperature and pressure increase rapidly: at a height of some 30 kilometers the temperature is already 200°C, and the pressure is more than 7 atmospheres; at ground level the temperature ranges from 460° to 480°C and the pressure is 90 atmospheres—about that of our oceans at a depth of one kilometer. (Thus were the *Venera 5* and *Venera 6* probes, designed to withstand pressures of up to 25 atmospheres, destroyed before they ever reached the ground.) With its density one-tenth that of water and its temperature 100°C hotter than that of molten lead, the physical conditions of the atmosphere near the surface of this planet are scarcely imaginable to us.

The high temperature results from a "greenhouse effect." (In a greenhouse the transparent glass admits light and heat but prevents the warmed air from escaping; thus the interior heats up.) Carbon dioxide is transparent to visible radiation, such as light, and opaque to infrared rays; therefore, when the solar radiation that reaches the ground is re-emitted in the form of infrared rays, it is blocked by carbon dioxide, and the heat builds up.

Even harder for us to imagine are the clouds. As had already been observed through the telescope, the cloud structure of Venus appears only in ultraviolet light, probably because of the inhomogeneous distribution in the atmosphere of a substance that absorbs ultraviolet rays. When observed in visible light, the outer atmospheric stratum appears yellow and uniform. Thus we are not dealing with aqueous clouds, for terrestrial clouds, formed of water vapor, are white, not yellow; moreover, as we have seen, water vapor is too scarce in the Venusian atmosphere to constitute a fundamental component of the clouds. It has been determined, from the distribution of the atmospheric temperature at various heights, from measurements of the polarization of the light reflected from the clouds, and from calculations taking these results into account, that the clouds of Venus are formed of very minute drops of sulphuric acid. More precisely, the clouds consist of an aqueous solution containing about 1.80% by weight of H_2SO_4.

There is reason to believe that as one descends to lower levels of the atmosphere, the number of drops per unit volume increases, until the droplets are sufficiently close to merge. Thus at a height of 80 kilometers, at the top of the cloud system, the solution is more or less in the form of a spray; at a height of 30–40 kilometers, a regular rain of sulphuric acid forms. In falling, the drops pass through an increasingly hot atmosphere; therefore, the water evaporates from their surface, and the acid becomes more and more concentrated; by the time the rain reaches the ground, the concentration must be around 98%.

Furthermore, it has been found that the atmosphere of Venus also contains—although only in small amounts—hydrogen chloride and hydrogen fluoride, which, reacting with the sulphuric acid, could form fluorosulphuric acid, a very strong acid capable of attacking and dissolving almost all common materials, including most rocks.

An atmosphere so deadly stirs our imaginations as we try to picture the terrain and the environment on the planet's surface. Unfortunately the clouds have always prevented direct observation. Nevertheless, with radar we have managed to make extensive, if rather poor, observations, and two probes have enabled us to photograph sufficiently well two very small areas of the surface. The radar observations disclosed rough and smooth regions, both high and low, and a recently obtained map reveals craters measuring up to 160 kilometers in diameter and 500 meters high. Overall, this research shows Venus to be a rather flat planet, much smoother than the moon or Mercury, with few variations in level and with very few elevations higher than a kilometer.

The first close view of the ground was provided by the two Russian probes *Venera 9* and *Venera 10,* which landed on 12 and 25 October 1975, respectively. Each took a panoramic photograph of the surrounding region (Fig. 2.5). Although the two probes landed 2,200 kilometers apart, the portions of the surface photographed appear similar. Both areas appear to be covered with rather flat stones, having a surface smooth enough to reflect the light of the sun, at least as well as the steel part of the probe that is seen in the center of the two photographs. The ground between one stone and the next is rather dark, indicating that the surface is uneven or covered with dust. The stones do not appear to be piled up but seem to be distributed uniformly over the surface, in a single layer, as if they had come not

Fig. 2.5 The ground on Venus photographed from nearby: *above,* from the *Venera 9* probe; *below,* from *Venera 10.* Note the strange shape of the rocks, the sharpness of the shadows, and the transparency of the atmosphere—characteristics that could not have been foreseen on the basis of our previous knowledge of the planet.

from geological outcrops but from a random rain of matter over the planet. Many stones appear rounded, perhaps as a result of the corrosive rains.

Because of the nature of the atmosphere, it was believed that the sun would not be visible from the surface of Venus; the sky was expected to be leaden, as dark as our sky on a gloomy day. But the objects visible in the two photographs cast sharp shadows, indicating that the solar disk was indeed visible from the surface, at least at the moment the photographs were taken. This finding, plus the fact that the sunlight reaching the ground is stronger than was expected, suggests that the atmosphere of Venus may still hold some surprises.

In regard to the wind, the *Venera 10* probe, like *Venera 8,* did not record at ground level the terrible rush of air present in the upper atmosphere but a light breeze of scarcely 12 kilometers per hour. This fact should not, however, be taken as an indication that beneath the deadly atmosphere lies a pleasant terrain. This "breeze" is in reality one of fire, fierce enough to split stones, as evidenced by some of those visible in the photographs. In fact *Venera 8* and *Venera 10,* although constructed specifically to operate in these conditions, survived the environment for only 53 and 65 minutes, respectively.

Thus whatever may still await discovery, the general picture is already impressive: flat, arid terrain, perhaps strewn with lakes of lava on which float half-melted rocks, like igneous icebergs; a fiery environment, bathed in red light because of the strong atmospheric absorption; the whole burnt by a fleeting sun larger than the one we see from our planet; a dark sky laden with clouds tens of kilometers thick, which let fall a corrosive rain. Moreover, the high density of the atmosphere and the strong up-and-down convective currents must produce striking phenomena of refraction and distortion, making the landscape, already monstrous, appear unreal and hallucinatory. Even Dante's hell is not so horrible: after all, it was conceived within human parameters.

But as we move out of the atmosphere, beyond the denser layers, once again we perceive the familiar starry sky. This sky appears more beautiful than ours, not only in contrast to the gloomy hell we have just left, but because it is adorned with a splendid bluish star, even brighter than Venus in our sky. It is Earth, our Earth, which we shall soon be passing on our way to Mars.

EARTH

As we proceed, moving away from Venus toward the outer regions of the solar system, this bright spark becomes increasingly brilliant, turns into a small blue disk, and grows steadily larger until we can begin to distinguish reddish or greenish zones in the midst of vast blue expanses and whitish stretches in the shape of wisps, streaks, and spirals, which slowly form, change, and dissolve. These are, respectively, continents, oceans, and clouds. The disk grows larger; we are able to distinguish an increasing number of details, and instinctively we look for our own country and the city where we live. But now that we have visited such inhospitable regions, now that we have begun to experience the loneliness of isolation in space, how could we possibly not define our "country" as the entire globe? Every part of it offers us air that we can breathe (even if it is not always wholesome), the necessary water (even if it is not always potable), and the company of millions of beings like ourselves (who, though we may not like them all, are still our brothers). Shall we ever find similar beings in any other part of space? Shall we ever discover a planet similar to our own? Who knows? For the time being, all we can see is that the earth, which we have reached and passed, is shrinking, becoming smaller, disappearing. And while we reflect that now we are really entering on the most adventurous part of our journey and that from now on we shall be moving farther away, at first toward the boundaries of the solar system, then toward the stars, right to the limits of the known universe, that luminous disk, where for an instant we thought we glimpsed our own house, shrinks rapidly, becoming just a bright spark in the sky.

MARS

But another luminary, this one red, comes to greet us. It is Mars, the most famous planet, the subject of so many pages of science fiction, and the focus of so many telescopes. As has long been known, it is a body smaller than the earth, with a diameter of 6,787 kilometers and a mass barely 0.11 times that of the earth. Its density is also less than the earth's (3.9 times that of water), and it moves around the sun at a mean distance of 227,900 kilometers.

Today, after eleven space probes have photographed and examined it, Mars turns out to be quite different from what we had imagined up until a few years ago, but it is of greater interest than ever and still so full of mysteries to be solved.

As we approach Mars, its brightness gradually increases until, at a certain distance from it, we begin to notice that the bright point is spreading into a disk. The disk steadily increases and finally reaches the size it had for the terrestrial observers of the last century when they followed it with their most powerful telescopes. We can now see plainly what they discovered only after long and patient research.

We note first of all that Mars shows bright and dark areas with well-defined outlines: the former are reddish, the latter gray. From the motion of these spots it is found that Mars rotates about an axis only slightly more inclined than that of the earth (23°59′) with a period of 24 hours, 37 minutes, and 23 seconds. The spectrograph shows the presence of an atmosphere more rarefied than that of the earth, and we can see two white caps in the polar regions, quite like those covering the terrestrial poles.

The inclination of the axis creates seasons similar to ours but somewhat longer, since, as Mars is farther from the sun, its year lasts 687 days, nearly twice as long as ours. If we were to stop and observe Mars from our space observatory for a couple of years, we should witness changes like those of the earth's surface as seen from the moon. If we observe a Martian hemisphere during its winter, the dark regions appear gray and the bright regions practically colorless, while the white polar cap attains its maximum extension (Fig. 2.6). With the coming of spring the polar cap steadily shrinks; the dark regions become more pronounced and take on a greenish cast, at first near the polar cap, then at ever greater distances, until the "green wave" reaches the equatorial zone. We have reached the beginning of summer; the polar cap has almost disappeared; the contrast between bright and dark regions reaches a maximum and then gradually lessens, until the cycle is complete; the process starts all over again with the new season. This cycle suggests that something is transported from the pole that primarily affects the dark regions: for example, water from the melting of the polar snows might revive some type of vegetation.

This picture, based on telescopic observations, seems simple and definitive: a planet like the earth, smaller, not as warm, with a more rarefied atmosphere, but living—that is, supporting some form of life, if only vegetable.

This picture still seems to be valid as we draw nearer to the planet, at least up to a distance of about a million kilometers, the distance from which *Mariner 6* and *Mariner 7* took the first photographs. From this distance Mars appears as it does observed from earth through a telescope, although the spots are easier to see and more detailed (Fig. 2.7). As we continue to approach Mars, let us take our eyes from the planet for a few hours and look in the opposite direction, toward the sun and the brilliant luminary shining beside it, the earth. Then, as we begin orbiting Mars from close by, let us turn to face it again—we cannot believe our eyes! The bright and dark areas have almost disappeared, and in their stead there appear, not shapes marking the same areas but a whole new mountainous region which seems to have no connection with the areas we observed not long ago.

We would think it another planet if we did not know that it is not easy to make such a large body disappear and make another suddenly appear in its place. Besides, something of the Mars we observed is still there—the polar caps, for example—and if we look carefully, we might recognize some other region as well. But as we continue to orbit the planet and observe more and more new regions, we forget our disbelief in our surprise over this completely new world being revealed to us, which is different not only from the earth, but also from the moon, Mercury, and Venus (Figs. 2.8a, b, c).

At a first glance we are struck by the regions abounding with craters, above all in the southern hemisphere. These craters have diameters ranging from a few kilometers to several hundred kilometers. Some are flat-bottomed; others are small and hollowed out like soup bowls (these are like the craters on the moon or Mercury, only flatter, more worn down).

We find ourselves beginning to think that, except for the earth, the planets are all alike—pitted with holes—when we notice an enormous flat region that extends over a large part of the northern hemisphere. This region, called Tharsis, is a kind of immense high plateau with a diameter of over 5,000 kilometers, its highest parts rising to 7,000 meters above the neighboring cratered terrain that we saw earlier. Here the impact structures are very few and small. The craters of the southern hemisphere are thought to have originated in a very remote epoch, over 4,000 million years ago, when fragments of the material left over after the formation of the planet crashed to the surface. If that is the case, then the high plateau in the north must have been formed later, when flowing magma obliterated the craters

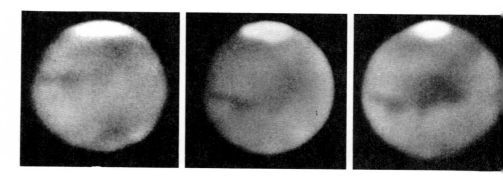

Fig. 2.6 A sequence showing the progressive change in the south polar cap of Mars and the spreading of the dark regions as the season advances. The phenomenon is described as the spreading of a dark wave from the pole toward the equator. The six photographs run from 10 March to 22 August (Martian calendar). (*Lowell Observatory*)

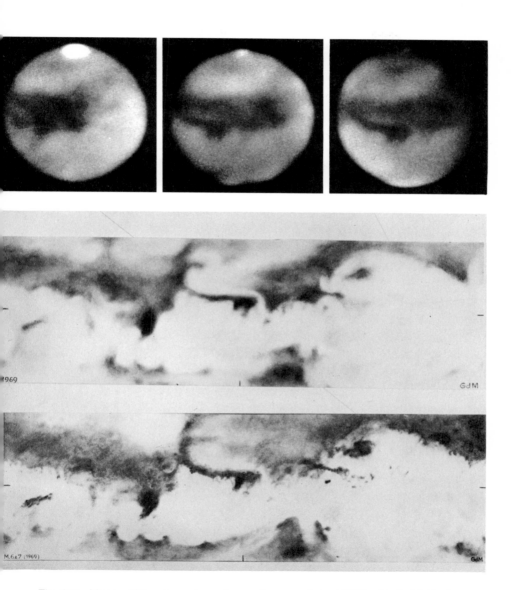

Fig. 2.7   Maps of Mars based on observations made in 1969 by G. de Mottoni y Palacios: *above,* compiled from photographs taken through a telescope on Earth; *below,* compiled from photographs taken by *Mariner 6* and *Mariner 7* as they were approaching the planet. Note that both give the same view, even though their resolution differs. On these maps and on the photographs in the preceding figure, north is below.

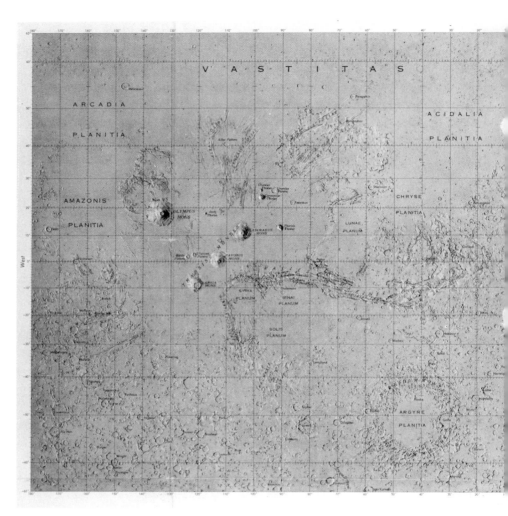

Fig. 2.8a   Map of the planet Mars derived from the thousands of photographs taken by the *Mariner 9* probe during its orbit around the planet. Note the general appearance, which is quite different from what was seen through the telescope. This map covers the area between the latitudes +60° and −60°.

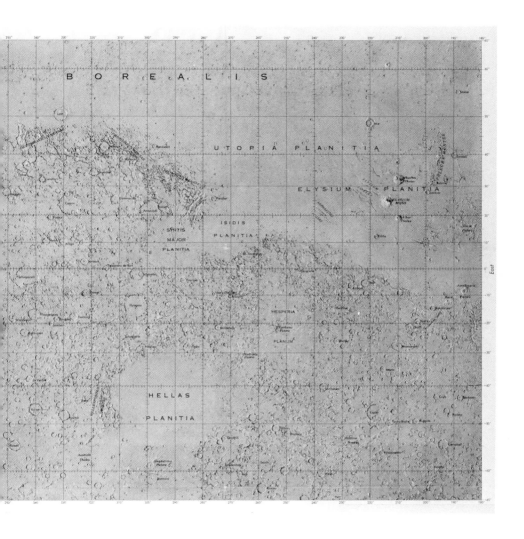

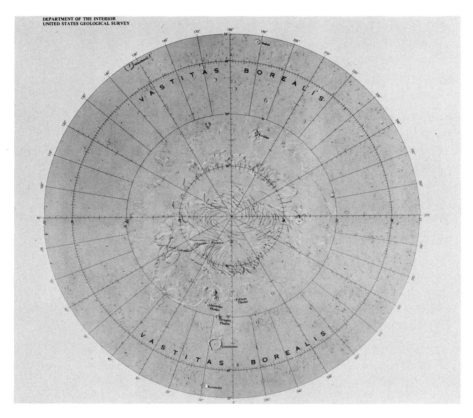

Fig. 2.8b   Map of the north polar cap of Mars compiled from *Mariner 9* photographs.

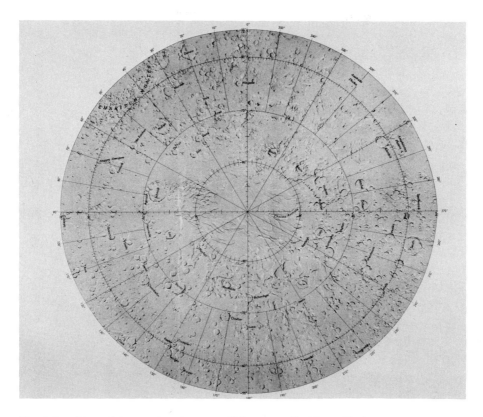

Fig. 2.8c   Map of the south polar cap of Mars compiled from *Mariner 9* photographs.

that, in the beginning, must have existed in the northern hemisphere as well.

On this immense plain rise four giant volcanoes, the biggest of which G. Schiaparelli had glimpsed through a telescope. Discovering a round, shiny white spot, the celebrated astronomer had attributed it to snow on a high mountain and called it Nix Olympica. Photographed at close quarters, this peak, snowy and sometimes covered in clouds, turns out to be the summit of an enormous volcano 26,000 meters high and 500–600 kilometers wide; near the base, it rises from the plain below in an almost vertical escarpment about 4,000 m high (Fig. 2.9). The peak, about 70 kilometers wide, contains a complex system of calderas, which must have been at one time the outlets for the lava flows. With its flattened profile, this volcano belongs to the type known as shield volcano, which is formed by successive eruptions of low-viscosity lava; capable of flowing great distances from the outlet, such lava forms escarpments that incline only slightly. The volcanoes of Hawaii are of this type; however, the largest of these, Mauna Loa, is only one-third the size of this giant, which has been named, in accord with Schiaparelli's nomenclature, Mons Olympus.

The other three volcanoes in the region of Tharsis have diameters of about 400 kilometers and are about 20,000 meters high. There are, moreover, various volcanic structures of a lesser diameter (about 100 km) and with much steeper slopes: they may have been formed from more viscous lavas. The region also contains many other volcanic formations shaped like saucers, each having a central caldera with a diameter about half that of the whole structure. In the more northerly part of this region we find the most extensive of the Martian volcanoes, Alba Patera, which must also be one of the oldest, since it has been almost entirely destroyed by erosion. The region of Elysium, which forms a bulge like that of Tharsis but is smaller and less elevated, also contains several volcanoes. The shield volcanoes are similar to those of Tharsis but are smaller and perhaps older. The biggest is Mons Elysium, 200–300 kilometers wide and about 15,000 meters high.

The southern hemisphere of Mars contains a few relics of volcanoes, pocked by impact craters. The quantity of such craters on their slopes shows that these volcanoes were active toward the end of the period when the planet was being heavily bombarded by the detritus remaining after the formation of the solar system. This would be about 3 billion years ago.

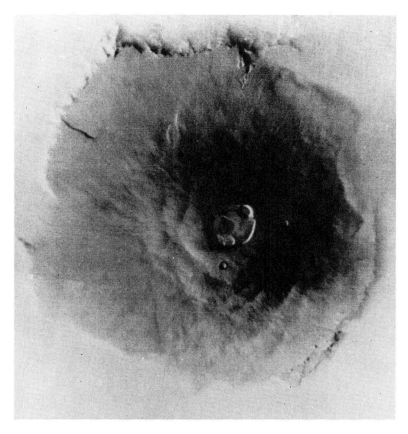

Fig. 2.9   The volcano Mons Olympus, photographed from *Mariner 9*. Its central crater has a diameter of 65 km, while the base has a diameter of 500 km and ends in steep escarpments. The height of the volcano above the surrounding plain is about 16,000 m. (*NASA*)

Thus, volcanic activity was already occurring on Mars at that time, and Mons Olympus, the most recently formed volcano, which is not more than 200 million years old (according to M. H. Carr), may still be active.

East of the three lesser volcanoes in Tharsis we see a striking formation: a kind of furrow seems to cut into the planet. This is the so-called Valles Marineris, a group of canyons 4,000 kilometers long, on the average 120 kilometers wide, and over 6,000 meters deep. This enormous formation, as long as the United States is wide, begins in the west with an intricate net of deep, narrow canyons and ends in the east in winding furrows like river-beds. These furrows, the longest of which stretch for more than 1,500 kilometers, are also found in various other regions of Mars and are very different from the lava channels of the earth and moon (Fig. 2.10). On the other hand, if we observe them more closely, we notice beds with a herringbone structure, unmistakable islands, and sometimes a system of tributaries very similar to those of our rivers (Fig. 2.11). Many contend that these beds were actually formed by water, but long ago, when the climate of Mars was hotter and more humid. Apart from these furrows, Valles Marineris shows no signs of having been hollowed out by erosion; according to the majority of scientists, it is of tectonic origin.

We have not yet exhausted the singular aspects of Mars. In the southern hemisphere, in the middle of a cratered terrain, there is an enormous circular area called Hellas, which is completely lacking in formations: it is a flat smooth stretch with no prominences whatsoever. This great basin, about 4,000 meters deep, must have been formed by the fall of a great meteorite during a very remote epoch; today it is perhaps the largest area of sand and dust on the entire planet. It makes sense that there should be a connection between Hellas and the great sandstorms that at times develop on Mars. Observed through the telescope, these storms were seen to be so violent and extensive that they would affect the entire planet, hindering observation of the underlying ground. Today it is known that they originate in the Noachis region, immediately northwest of the Hellas basin. They arise suddenly each year at the start of the southern summer; we are of course referring to the Martian "year" and "summer." One of the greatest storms, in 1971, was followed for a long time from the earth and subsequently, toward the end, from the *Mariner 9* probe. The phenomenon appears, above Noachis, as a lengthy, shining white streak, with a front of some thousands of kilometers.

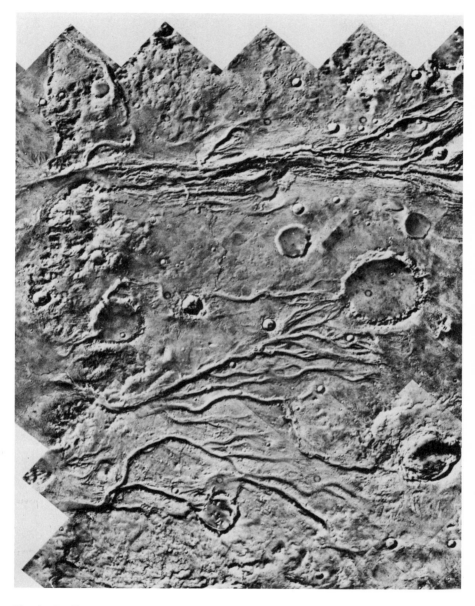

Fig. 2.10   Furrows like river beds between Lunae Planum and Chryse Planitia, on a terrain sloping from left to right, the difference of level being about 3 km. Some furrows, like the system above, have obliterated preexisting craters; others, like that from center to right, debouch into craters. It is thought that these furrows were fashioned by water, but the original source, in the southern zone of Lunae Planum, has not yet been identified.

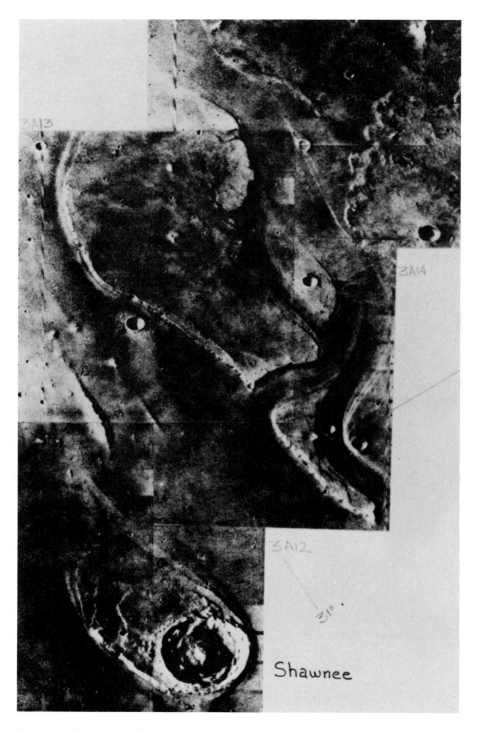

Fig. 2.11   Formations like river beds and islands in a river, photographed by *Viking 1* as it circled the planet.

The dust cloud is swept along by strong winds having a speed estimated to be at least 150 kilometers/hour. The central region of the dust cloud spreads at first very slowly, then with increasing speed; it expands primarily westward, and ultimately affects the entire planet. As it spreads, it branches north and south, covering the whole southern hemisphere in a little less than a month. At times, as happened in 1971, it also extends to the northern hemisphere and the entire planet is smothered under a cloud of sand and dust for a period of up to four months. Eventually the storm abates, the dust particles fall, and in a few months the atmosphere regains its normal transparency.

The observation of the atmosphere is very important if we are to consider the most interesting aspect of Mars: its habitability. Unfortunately the information we have in this respect is not encouraging. The Martian atmosphere is very rarefied: the pressure at ground level is roughly equal to that of our atmosphere at the height of 35,000 meters. Furthermore, as was confirmed by the *Viking 1* probe in July 1976, it is composed almost exclusively of carbon dioxide. Carbonic acid represents 95%, nitrogen 3%; the remaining 2% is made up of argon and traces of other gases. Water vapor is extremely rare, hardly 0.01%, though this fact does not necessarily preclude the presence of water on Mars.

The white polar caps, once believed to consist exclusively of water ice, must also contain dry ice, that is, carbonic acid ice. It is thought that once a year one-quarter to one-sixth of the atmospheric carbon dioxide condenses on the pole of the winter hemisphere, to evaporate in spring and condense once again on the opposite hemisphere. In this way the polar caps, which descend to a latitude of about 60°, are formed. The terrain of the polar regions is made up of sedimentary deposits of two types: one, which is older and not stratified, covers a plain pitted with craters, and extends to lower latitudes; the other, nearer the pole, is formed of a series of sheets each about 30 meters thick, decreasing in size from the bottom upward (Fig. 2.12). Actually, both of these deposits are eroded by the melting ice. The material carried away is redeposited in the middle latitudes, where it forms a thin surface layer and partially fills the interior of the old craters.

When, in summer, a cap melts, it does not disappear completely: a small cap remains that never dissolves. This cap must consist exclusively of

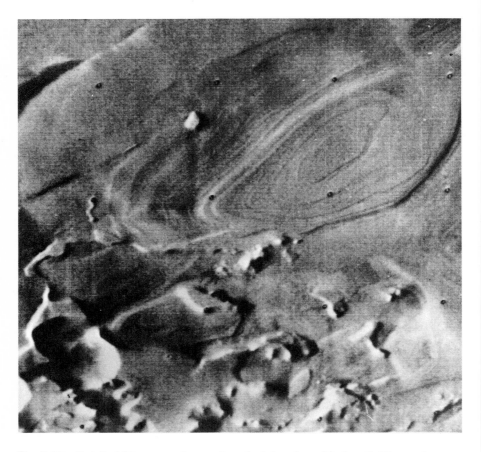

Fig. 2.12 Detail of Mars near the south pole, taken from *Mariner 9*. The oval-shaped contours are perhaps due to superimposed strata of dust, volcanic ashes, and ice. (*NASA*)

water ice, which melts at a higher temperature than the melting point of dry ice. A certain quantity of water is certainly present on the highest mountain peaks and is perhaps imprisoned at all latitudes in the sand scattered over the rocks. Furthermore, at latitudes above 45° the ground under the surface must be cold enough throughout the year to have caused a great part of the water vapor present in the atmosphere to condense in layers of subterranean ice analogous to the permafrost present on our planet in the regions near the poles.

Thus it now seems certain that the atmosphere of Mars is poor in water vapor because all the water in the planet is concentrated in the polar caps, in the permafrost, and in the sand. It has been calculated that if all the water vapor present in the atmosphere were to condense, it would cover the surface of the planet with a skin a few hundredths of a millimeter thick. But if, instead, the water in the polar caps were distributed over the whole planet, it would form a layer about ten meters deep. An equal volume, if current theories are correct, would be obtained by melting the water present in the permafrost and in the sand. This fact is attributable to the low temperatures that prevail on Mars, even in the warmer equatorial zones. It seems that even in the more temperate zones the temperature ranges from +15°C at midday in summer to −100°C on a winter's night. The two Viking probes, which landed on the planet on 20 July and 4 September 1976, respectively, have provided this kind of precise information, although it is drawn from only two localities (Figs. 2.13, 2.14).

At this point let us also stand on that ground, to touch it, look around us, and raise our eyes to another sky. Let us land on a spot in an intermediate latitude in the northern hemisphere, like the site where *Viking 1* landed, toward the end of the Martian spring. Apart from the fact that the atmosphere, very thin and different from ours, is not breathable, the environment is not very different from some terrestrial environments: dry and cold, descending to −40°C and even by day, in sunlight, never exceeding −5°C; almost no wind (at most not more than 18 km/h); clear sky. Equipped with a respirator and winter clothing, we could survive under these conditions, and that is saying a lot, when we consider the infernal planets, Mercury and Venus, that we visited previously. The landscape is that of a desert: sand dunes alternating with scattered rocks. But it is a strange desert, unlike any

Fig. 2.13 Panoramic view of the Martian landscape, over an arc of 100°, from the spot where *Viking 1* landed. The landscape is desertlike in appearance; its sand dunes are produced by a wind blowing recently from left to right and it is strewn with rugged rocks. The boulder on the left, 1 by 3 m, is about 8 m away. The sun, which rose a couple of hours earlier in an almost cloudless sky, shines on this ground that is devoid of life.

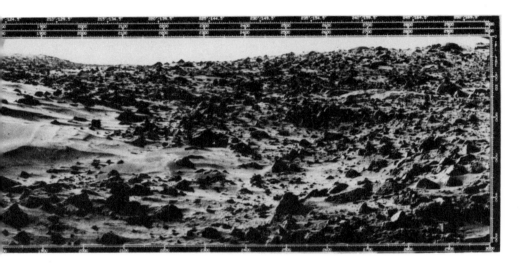

Fig. 2.14 Strangely pitted rocks in the landing area of *Viking 2,* visible, as far as the eye can see, to the horizon about 3 km away. The nearest rock (lower right) is about 25 cm in diameter. In the foreground we see a sort of dry rivulet meandering from left to right. The line of the horizon appears sloping because when the probe landed, one of its feet came down on a rock.

found on our planet, for it is a rusty red, and a good part of the surface material must be iron oxide. The sky too is red, or pink, perhaps because of the very fine dust particles stirred up by the wind.

Much more like our own is the landscape at night. When, after sunset, the colors disappear and objects take on the same grayness as terrestrial objects seen at night, the red desert is still, and in the night sky the stars light up, one by one. We find the same constellations for, in going from Earth to Mars, we have taken only a very tiny step in the solar system and one quite imperceptible in relation to the stars, which are enormously more distant.

All the same, there is something different in this sky: the most obvious difference is the presence of two satellites—two moons instead of one. They revolve fairly close to the planet, one at a distance of 9,300 kilometers and the other at 23,300 kilometers; but even though they are so close, they are so small that the closer one appears only one-third as large as the moon seen from earth, and the other is just a large bright spot. Moreover, they move in the sky in a strange way. The more distant satellite, which makes a complete revolution around Mars in 30 hours and 18 minutes, rises in the east and sets in the west like other heavenly bodies, but it appears to move extremely slowly; from any given point it remains above the horizon for fully 64 hours. The nearer one, on the other hand, makes a complete revolution around Mars in 7 hours and 39 minutes, that is, in much less time than it takes Mars to turn on its axis. Thus it will be seen to rise in the west and set in the east, having crossed the sky in about 5.5 hours.

As the night advances, we notice that the constellations do not move around the pole star, as they do when observed from earth. The axis of rotation of Mars is in fact directed toward a region between the constellations of Cepheus and Cygnus that lacks bright stars. What we see is similar to the sky viewed from the southern hemisphere of Earth. As the night progresses, we also see a few planets. Jupiter, Saturn, and the more distant planets appear as they do when seen from Earth. It is difficult to pick out Mercury; Venus, however, is clearly discernible in the morning or evening sky. And finally, as the first light of dawn appears, another star, a little less bright than Venus in our sky, rises over the Martian desert, which emerges from the darkness of night and resumes its reddish color. This new body, which does not appear in our sky, is the earth, which, rising over the horizon, tells us that a new day is beginning.

With this sight we leave Mars, a planet so different from our own and yet sufficiently similar for us to survive there, perhaps even comfortably. We leave it to head toward the outer regions of the solar system, where we shall encounter giant planets that only recently have begun to disclose their mysteries to us.

THE MINOR PLANETS

One of the most interesting discoveries made by *Mariner 4,* the space probe launched toward Mars in 1965, was that the number of micrometeorites increases remarkably as one moves away from the earth in the direction of Mars. In fact, whereas *Mariner 2,* which was launched toward Venus (that is, in the opposite direction), recorded the impact of only two meteorites, *Mariner 4* recorded 77 impacts in the first 3,100 hours of travel, and subsequently the number of particles that collided with the probe increased to such an extent that NASA technicians began to fear a collision with a body large enough to damage or destroy the vehicle, thus jeopardizing the success of the undertaking.

The presence of these particles did not come as a surprise, for as it was nearing Mars, *Mariner 4* was also approaching the so-called asteroid belt, a vast zone lying between the orbit of Mars and that of Jupiter, within which an indefinite number of planets of miniscule dimensions circle the sun (Fig. 2.15).

The first of these planets was discovered by chance on the night of 1 January 1801 by G. Piazzi, director of the Palermo observatory. The astronomers of the time reacted to news of the discovery with joy and a certain degree of relief. The new planet (which its discoverer named Ceres) had been sought for some time. In fact, according to a certain empirical law, known as Bode's law, there should have been a planet at a distance of about 418 million kilometers from the sun. None of the planets known since antiquity fulfilled this condition; according to Bode's law, Mars occupied the fourth place from the sun, Jupiter the sixth, and the fifth remained vacant. Now Ceres very conveniently occupied this fifth place.

The astronomers' joy was, however, tempered when on 28 March 1802 W. Olbers discovered another planet, with an orbit similar to that of Ceres; it was named Pallas. On 1 September 1804 a third planet, Juno, joined the

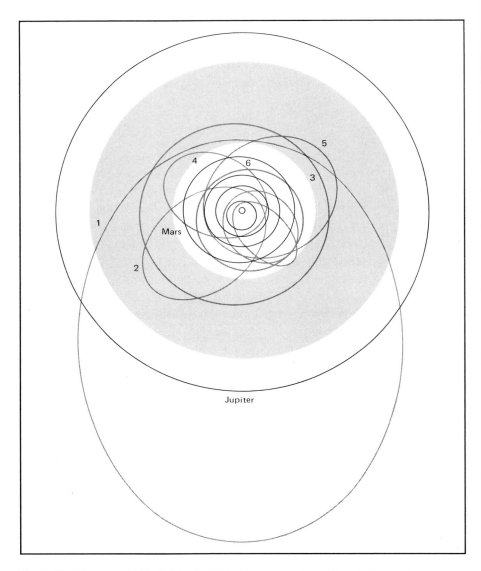

Fig. 2.15   The asteroid belt (shaded) lying between the orbits of Mars and Jupiter. The numbers indicate those asteroids which follow eccentric orbits: 1, Hidalgo; 2, Adonis; 3, Ceres; 4, Apollo; 5, Amor; 6, Eros; 7, Icarus. The almost circular orbits of the planets are shown as darker lines.

others, and on 29 March 1807 the brightest of the group, Vesta, was discovered. Since then the number of bodies of this type known to us has been
constantly increasing. In 1868, 100 were known; in 1879, 200; in 1890, 300.
At the end of 1891 the Heidelberg astronomer M. Wolf adopted the photographic method for this research, and the discoveries mounted: by the end
of 1974 fully 1,914 had received definitive numbers. The majority of these
bodies may still be awaiting discovery. According to a recent statistic, based
on the increase in their number as smaller and smaller bodies are included,
there should be some 22 million, and even thousands of millions, when the
smallest fragments are taken into account. This estimate would be very
difficult to validate, for obvious practical reasons, but the Leyden and Mount
Palomar observatories have been conducting a search for the fainter bodies
and have discovered several thousand; the orbits of many of these bodies
are now being studied, and their definitive numeration is being determined.

These heavenly bodies are called minor planets, or asteroids, because
they revolve around the sun as do the larger planets but are of greatly
reduced dimensions (Fig. 2.16). Furthermore, they do not in fact resemble
their larger counterparts. The largest, Ceres, has a diameter of scarcely 955
km; Pallas, Vesta, and Juno are smaller still. Only some thirty have a
diameter exceeding 100 km; most of the others are less than 50 km, and
those discovered more recently are actually less than 20 km.

It seems that only Ceres, Pallas, and Vesta are perfectly round. Most of
the others show periodical variations in brightness, which clearly indicate
the presence of an axial rotation and some irregularity in shape; this effect
may also be due, at least in part, to the alternation of bright and dark zones,
reflecting more or less sunlight. It was possible to determine more precisely
the appearance and dimensions of Eros, one of the better-observed asteroids when on 23 January 1975 it passed in front of a star. Eros turned
out to be a disk about 19 by 30 kilometers and 7 kilometers thick. On such a
celestial raft, instead of a horizon there are actually two edges, each some
hundred kilometers long, which one can lean over to survey the sky below.

Because of their small size and consequently low gravitational fields,
none of these asteroids could possibly retain an atmosphere. Moreover, the
absence of an atmosphere has been confirmed by spectroscopic observations. In short, what we are dealing with here, for the most part, are simply
large rocks orbiting in space.

It was stated that the minor planets revolve around the sun in orbits lying between those of Mars and Jupiter. This is true only for most of them and only as a rough approximation. In reality all orbits, for the minor as well as the major planets, are at least slightly elliptical. In fact some of the minor planets have highly elliptical orbits that bring them, at least for part of their revolution, much nearer to the sun than Mars or even the earth ever gets. One of these minor planets actually comes within 28 million kilometers of the sun, closer even than Mercury; thus it ought to be considered the planet closest to the sun. This venturesome object, discovered in 1949 by W. Baade, has been given the name Icarus, after the legendary son of Daedalus who lost his life because he dared to approach too near the sun. (Baade also discovered, in 1920, the minor planet that moves farthest away from the sun: it is Hidalgo, whose orbit reaches out almost to that of Saturn.)

Icarus can also come very close to the earth: on 15 June 1968 the two bodies were only 6.7 million km apart. Icarus, however, is not the asteroid that comes closest to the earth. Among those that honor us in this way are Apollo, which in 1932 was 3.2 million km distant; Adonis, which in 1936 was only about one million km away; and finally Hermes, which in 1937 came within 780,000 km, about twice the distance between us and the moon. So far about twenty asteroids of this type are known.

The fact that some asteroids come appreciably close to the earth might give rise to the suspicion that a small one, as yet undiscovered, could actually collide with our planet. Such an occurrence, although extremely improbable, is not impossible. And this possibility shows the practical importance of studying the asteroids. The timely discovery of such a body would allow us to predict where it would fall; we could even, perhaps, try to destroy it in space with atomic weapons. All the same, we should not place too much faith in the efficacy of the warning system. It is enough to recall that on 10 August 1972 a body about ten meters in diameter and weighing at least a thousand tons penetrated our atmosphere without warning and passed through it without falling. If its orbit had passed even a few kilometers closer to the ground, it would certainly have fallen in the Canadian province of Alberta, causing an enormous explosion, creating a large crater, and spreading death and destruction.

Some astronomers have proposed that asteroids be used as probes in exploring the solar system. It would be easy enough, for example, to take advantage of one of the periodical approaches of Icarus in order to install an unmanned station on it; this station would then be transported through the solar system, from the distance of Mars to the immediate vicinity of the sun, and could transmit some very interesting observations of the sun and interplanetary space at various distances from us.

From the theoretical point of view, perhaps the most interesting question in regard to the asteroids concerns their very existence. Indeed, it is not yet understood why, instead of a major planet, we have innumerable insignificant bodies. Naturally, various theories have been proposed in this connection, and when we succeed in determining the correct one, perhaps we will be able to clear up the mystery of the origin of the entire solar system.

THE GIANT PLANETS

JUPITER

When we leave the zone of the asteroids, the Lilliputians of the solar system, and head for the planets farther from the sun, the first that we reach is Jupiter, the largest known planet (Fig. 2.17).

It revolves around the sun at a mean distance of 778 million km, making a complete revolution every 11 yr, 317 d. Though it weighs 318 times as much as the earth, its density is barely 1.33 g/cm^3. It has an equatorial diameter of 142,800 km and is so flattened at the poles that the effect can be observed through even a small telescope. Such a telescope would also show us that Jupiter's disk appears to be furrowed by alternating bright and dark bands, or belts, parallel to the equator. A somewhat larger telescope will show certain structural features that can serve as points of reference for determining the planet's velocity of rotation around its polar axis. It is thus found to make a complete rotation every 9 h, 50 m, 30 s; but the velocity is not the same at all latitudes because, as is the case with Venus, what we are observing is only the upper layer of the atmosphere.

Nonetheless, many of the observed configurations seem fairly permanent. Let us take as an example the most famous of these, the so-called red spot. Discovered by G. D. Cassini in August 1665 and observed by him and others for over half a century, by the beginning of the eighteenth century it was universally known and, indeed, considered a characteristic so typical of

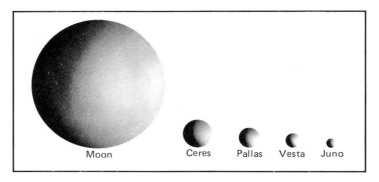

Fig. 2.16   Dimensions of the asteroids Ceres, Pallas, Vesta, and Juno compared
with the Moon.

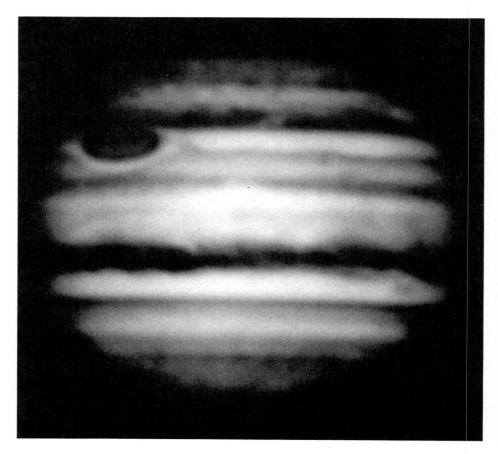

Fig. 2.17   The planet Jupiter, photographed in blue light with the large telescope of
Mount Palomar Observatory on 24 September 1952. Note the polar flattening, the
darkening toward the limb (as for the sun) on account of the thick atmosphere, and
the characteristic dark bands parallel to the equator. The southern hemisphere is
dominated by the famous oval-shaped red spot. (*Mount Wilson and Palomar Observatories*)

the planet that it is shown in a 1711 painting by Donati Creti now in the Vatican. Later on, this configuration faded to such an extent that it was completely forgotten; when the astronomer H. S. Pritchett observed it anew in July 1878, he thought it was being seen for the first time. From that time on, it has been followed and studied continuously; today we can draw a picture of it that is reasonably complete and quite interesting.

The red spot is a formation elliptical in appearance and brick red in color. The shade and intensity of the color vary with time, to the extent that at certain epochs the spot actually disappears. Its dimensions are 13,000 by 40,000 kilometers; thus it could easily contain three globes the size of the earth. Its latitude is nearly constant, but the longitude varies, that is, the velocity of rotation of the spot is not the same as that of the rest of the planet. Thus, although it must be consistent and permanent (as is proved by the fact that it has been in existence for at least three centuries), it seems to float over the rest of the planet.

As we have already seen, Jupiter shows alternate bright and dark belts, parallel to the equator. The red spot is on a bright belt, but the underlying dark belt is curved along its boundary; that is, it follows the edge of the spot. There is also, in the southern hemisphere, at a slightly different latitude from that of the spot, a group of configurations known as the great southern disturbance. Because of their different speeds of rotation, the red spot and the disturbance occasionally meet. As the disturbance approaches the red spot, it deviates toward the edge of the latter and, accelerating, skirts alongside it; then it gradually slows down again as it moves away.

It was recently discovered that the red spot is not unique. In 1972 a small spot appeared in the northern hemisphere, and when, eighteen months later, the *Pioneer 10* probe photographed it from nearby, it was seen to be similar in shape and color to the great red spot; the only difference was that the dimensions were greatly reduced. When the planet was photographed again the following year by *Pioneer 11*, the small red spot had disappeared. It is thought to have lasted for about two years. The appearance of this second spot raises a question: Can we be certain that the red spot discovered by Cassini and that found by Pritchett are the same spot?

These phenomena seem very strange to those accustomed to picturing the planets as bodies with a well-defined solid surface, like the earth or Mars. Jupiter must be quite different. The formations on Jupiter appear well-defined and, on the whole, stable. Thus they cannot be attributed to a constantly varying atmosphere of the terrestrial type. Neither can they be attributed to the surface, for they do not rotate all together like a rigid body; moreover, they vary constantly in so many details.

The *Pioneer 10* and *Pioneer 11* probes (which passed close to Jupiter in early December 1973 and 1974), laboratory research, and theoretical calculations allow us to interpret the visible part of Jupiter thoroughly and to form a very plausible model for the underlying zones. The conclusion reached is that what we see is certainly the outer part of an atmosphere some 1,000 kilometers thick, composed of hydrogen, helium, ammonia, methane, and water. Infrared observations made by the two probes have shown that the bright belts are colder and higher than the dark belts and, furthermore, that the gases in the bright belts are ascending, while those in the dark belts are descending. Landing on the planet, we should encounter hydrogen gas first of all, then the higher clouds made up of crystals of frozen ammonia. Subsequently we should pass through a layer of crystallized ammonium hydrosulphite (which gives the dark belts their characteristic color), a layer of ice crystals, and finally a layer of water droplets. Below this last we should finally touch ground, but here awaits the greatest surprise of all: there is no ground.

Relying on cosmogonical considerations, on the gravitational analysis carried out by the two *Pioneer* probes, on the magnetic field, and on the extrapolation of laboratory data relating to the behavior of hydrogen at the high temperatures and pressures that must prevail on Jupiter, J. D. Anderson and W. B. Hubbard succeeded in presenting a very convincing model of the planet's interior. According to this model, we should encounter below the atmosphere a liquid surface: a single ocean of liquid molecular hydrogen about 24,000 km deep, covering the entire planet. At the bottom of this ocean we find a temperature of 11,000°C and a pressure of 3 million atmospheres. As we descend further, the hydrogen can no longer maintain itself in the molecular state; it passes to the liquid metallic state, which cannot be observed in the laboratory, because it can exist only at extremely high pressures. In this state the hydrogen dissociates into atoms, forming a fluid that is an excellent conductor of electricity. This layer, 43,000 km deep, perhaps accounts for the strong, complex magnetic field found by the *Pioneer* probes around the planet. Finally, in the central zone we find a nucleus with a radius of scarcely 3,000 km, composed of iron and silicates (the principal constituents of the earth) at a temperature of about 30,000°C.

In sketching this picture, the scientists have also taken into account an unexpected discovery made by *Pioneer 10*: Jupiter emits about twice the heat it receives from the sun. Thus, Jupiter has an internal source of heat and, unlike the other planets we have encountered so far, which merely reflect the light and heat they receive from the central body, behaves like a miniature sun, a sun whose center is less than one-thousandth as hot as that of the sun around which it revolves, but is all the same ten thousand times hotter than any terrestrial furnace.

Such, then, is the basic structure of Jupiter—perhaps. Such also, apart from the proportions, appears to be the structure of Saturn—the planet farthest from the sun of those planets visible to the naked eye and known from antiquity.

SATURN

Saturn revolves around the sun at a mean distance of 1,428 million km; it covers its orbit in 29 y, 167 d. Its mass is 94.15 times that of the earth, but its density is scarcely 0.69 g/cm³. It is smaller than Jupiter (its equatorial diameter is 120,800 km) and, like its larger counterpart, is greatly flattened at the poles. Direct observation shows that Saturn's disk also is crossed by belts similar to those of Jupiter, and the spectroscope has shown that its atmosphere has the same basic chemical composition: hydrogen, helium, ammonia, and methane.

The axial rotation of Saturn takes place in 10 h, 14 min, but, again, this value refers to the equatorial zone, since at other latitudes the speed varies.

Despite its similarity to Jupiter, Saturn is distinguished from every other planet in the solar system by one unusual feature: its famous, extremely beautiful rings. [Uranus was recently discovered to have rings.]

The earliest observations go back to 1610, when a ring appeared to Galileo in the form of two small stars on either side of the planet. The stars disappeared in 1612 and reappeared in 1616. Galileo gave up following so eccentric a planet, but various other astronomers, using improved telescopes, continued to draw these strange appearances, and even the ring as such, without ever recognizing it for what it was.

The mystery was cleared up by C. Huyghens in 1659. A few years later

Cassini discovered a gap—there were two rings. A third ring was discovered by W. C. Bond in 1850 and a fourth by P. Guérin in 1969. Thus today we speak of the *rings* of Saturn.

From studies made in the nineteenth century, it was found that Saturn's rings could not be rigid bodies; they are, rather, swarms of meteoric particles, revolving around the planet like so many satellites. According to very recent observations, the particles appear to be covered with snow or even fragments of ice and to have diameters of 4 to 30 cm.

The maximum width of the rings is 276,000 km; they approach to within 11,000 km of Saturn's surface. The thickness is minimal: it is thought to be no more than 20 km, and according to observations made by Dollfus in 1966, it is in fact scarcely 2.8 km.

Though minute, this thickness can be estimated, owing to a particular circumstance. Because of the various positions that the earth can occupy relative to the plane in which the rings lie, we see the rings sometimes from one side, sometimes from the other, and, fleetingly, in profile (Fig. 2.18). No wonder Galileo noted so many strange things about the planet Saturn. When he observed it in 1610, the two extreme ends appeared, in his small telescope, like two small stars; in 1612 the rings appeared in profile, and he no longer saw anything; finally, in 1616, conditions being once more favorable for observing the rings, he again noticed the two stars.

The disappearance of the rings occurs at intervals of 15 y, 9 mo; 13 y, 8 mo; 15 y, 9 mo; and so on. With modern telescopes we are able to follow the rings until they seem to be reduced to an extremely fine thread, immediately before or after they disappear totally. It has thus been possible to estimate their thickness, which remains the same over the whole plane, except at some points where a slight thickening is seen.

Even if observed through a small telescope, Saturn's rings are an unforgettable sight. Yet this sight is hardly stunning when compared to the spectacle that would entertain us from Saturn itself. Those immense arches, with their various gradations of color, luminosity, transparency, reaching up to the sky and partly interrupted by the shadow of the planet, must be awesome and marvelous. The lighting conditions change in the course of Saturn's year, and to an observer traveling on Saturn from the polar regions to the equator, the rings' appearance would change (just as they show various aspects before disappearing completely). This stupendous sight has probably never been witnessed by any living being—nor is it likely to be.

URANUS

If this voyage we are making had been undertaken only two centuries earlier, when we arrived at this point we would have stopped, certain that we had reached the bounds of the solar system. Indeed, up to the eighteenth century, the known planets were Mercury, Venus, Mars, Jupiter, and Saturn. Since then three other planets beyond Saturn have been discovered—Uranus, Neptune, and Pluto—and we are about to explore them.

Uranus was accidentally discovered by W. Herschel on the night of 13 March 1781. Other astronomers had observed it since the late seventeenth century, mistaking it for a star. Herschel, using a more powerful telescope, noticed that this new "star" showed an appreciable diameter; that is, it did not appear pointlike as all the stars do. On successive nights he recorded its apparent displacement in the sky, from which other astronomers derived its actual orbit in space, concluding that it was a planet more distant than Saturn.

Subsequent observations have shown that Uranus completes a revolution around the sun every 84 y, 7 d, at a mean distance of 2,872 million km. It has a diameter of 51,000 km, a mass 14.5 times that of the earth, and a density 1.2 times that of water. Although it appears very faint, Uranus is sometimes visible to the naked eye and could even have been discovered by the ancient astronomers without the aid of telescopes. Nevertheless, it is so distant that even the most powerful telescopes can tell us very little about its physical nature. Under optimal conditions of visibility it appears as a small greenish-blue disk; faint belts parallel to the equator, like those of Jupiter and Saturn, have been glimpsed (although photographs taken in 1970 through a telescope mounted on board a stratosphere balloon did not confirm the presence of these belts). It is flattened at the poles, a little more than Jupiter and a little less than Saturn, and its velocity of axial rotation is rather high, its period of axial rotation being 10 h, 49 min, which is the same order of magnitude as that of Jupiter and Saturn. Uranus, however, has one peculiar feature: the polar axis around which it rotates lies almost in the plane of its orbit, so that Uranus alternately turns one pole toward the sun for a very long period while the opposite pole remains dark and cold.

Uranus also has an atmosphere, which spectroscopic analysis has shown to be similar to that of Jupiter and Saturn: hydrogen, helium, and methane, which perhaps form actual clouds. It is certainly not an ideal place in which to live, and perhaps, as is the case with Jupiter and Saturn, we should not find that Uranus has ground on which to set foot. That fact does not prevent us from imagining a short stay there in order to glance at space from that planet, so close to the limits of the solar system.

Let us examine first of all the starry sky. We are 2,700 million kilometers from the earth—it would take a light ray, traveling at the speed of 300,000 kilometers per second, 2½ hours to cover this distance. If we wished to send a message to a base on Earth by the fastest means available (electromagnetic waves), we should have to wait 5 hours to get the reply. Yet our displacement in space has been so imperceptible, compared with the distances to the stars, that their apparent positions in the sky remain practically unchanged. The constellations and indeed the starry sky appear exactly the way they appeared from Earth.

Quite different, however, is the appearance of the solar system. Only three planets are still visible to the naked eye. Mercury, Venus, Earth, and Mars have disappeared in the immediate vicinity of the sun, completely overwhelmed by its light. Jupiter and Saturn appear at times as morning or evening stars; they never move more than 16° and 33°, respectively, from the sun. Jupiter appears as a star of at most magnitude 5, Saturn as one of 4; thus both are very faint. Neptune, the more distant planet that we shall soon visit, is visible, though extremely faint, only during the few years when it is closest to Uranus; then it disappears for about 150 years. Pluto is invisible even in the most favorable conditions.

Thus as we look out from Uranus, the solar system has almost entirely disappeared—it fades in the distance or is subsumed by the light of the sun. And the sun itself appears strikingly diminished. Its apparent diameter is barely 100″, and we see it no longer as a disk, but as a point—an extremely bright point, twelve hundred times brighter than the full moon as seen from the earth and thirty million times brighter than Sirius, the brightest of the stars. However, we have not yet reached the limits of this desolation and isolation, for at a distance of 4,498 million kilometers from the sun we find another planet, Neptune, which makes a complete revolution around the sun every 164 years, 280 days of terrestrial time.

NEPTUNE

U. Le Verrier discovered Neptune without ever having seen it; it is said, in fact, that he never saw it in his lifetime. He deduced its existence from his study of the perturbations of Uranus, which did not travel along its orbit according to the calculations of astronomers but behaved as if an invisible body were attracting it in such a way as to modify its path in space. From the differences between the observed and calculated data Le Verrier determined the weight of the perturbing body, calculated its orbit, and indicated the exact point in the sky where it should be sought. On the evening of 23 September 1846 the astronomer J. G. Galle of Berlin, observing the zone indicated by Le Verrier, found the new planet.

Neptune's diameter (49,500 km) is slightly less than that of Uranus, its mass is 17 times that of the earth, and its density is 1.7 times that of water. Telescopic observation, even more difficult than for Uranus, shows barely perceptible shadows that do not seem to have the appearance of the usual belts parallel to the equator observed on Jupiter, Saturn, and—perhaps—Uranus. The period of rotation, determined spectroscopically, seems to be 16 hours. The atmosphere, which appears to extend to a height of about 500 kilometers, is found to be composed essentially of methane, hydrogen, and helium. But we do not know to what extent we may speak of an "atmosphere" with respect to a body so far from the sun that its surface temperature must be less than $-200°C$!

THE FARTHEST PLANET: PLUTO

Beyond Neptune, at a mean distance from the sun of 5,910 million kilometers, revolves another planet, Pluto (Fig. 2.19). It was discovered by Clyde Tombaugh on 18 February 1930, in a position very close to that predicted, from calculation of the residual perturbations of Uranus, by the astronomer Percival Lowell, who had died fifteen years earlier. The situation would appear to be similar to that for the discovery of Neptune, but it has been shown recently that the discovery was made only because of a fortunate combination of circumstances.

The orbit calculated by Lowell was valid for a planet with a mass 6.7 times that of the earth. The new planet appeared at once too small and too faint to have this mass. In fact measurements of the diameter, carried out in 1950 by G. P. Kuiper and M. Humason with the world's largest telescope, gave a value of 5,800 kilometers. If one were to accept this value and even to assume that the mass is only equal to that of the earth, the density would turn out to be 10 times that of our planet—that is to say, 2.5 times that of gold! Since it seemed unreasonable to accept a density so different from all the matter forming our earth and the other planets under normal conditions, astronomers concluded that the value assigned to the mass or that determined for the diameter must be erroneous.

The latter datum, however, was then confirmed by an unusual observation. During the night of 28–29 April 1965 Pluto was due to occult a star in the constellation Leo. The phenomenon was followed in many observatories, but the occultation did not take place. Since it was found that the center of the planet had passed 0″.125 south of the star, it was deduced that the diameter of the planet could not exceed 6,400 kilometers, which agreed with the measurements obtained by Kuiper and Humason.

At this juncture three astronomers of the U. S. Naval Observatory reexamined all the observations of Neptune since its discovery and added 158 observations made in their own observatory between 1960 and 1968. Having also reexamined the values of the masses of Saturn, Uranus, and Neptune, they finally came to the conclusion that Pluto has a mass hardly 0.11 times that of the earth. If we assume the diameter is 6,400 kilometers, the density is found to be equal to 0.88 that of the earth, or 4.86 times that of water—a perfectly reasonable value. In other words, in terms of mass, diameter, and density, Pluto appears very similar to Mars. And if the mass is little more than one-tenth that of the earth, Lowell's orbit cannot be considered valid.

How then did C. Tombaugh, in February 1930, find it in a position not far from the one it should have occupied according to the orbit? I repeat: By chance. It was a fortuitous accident, like that leading to the invention of the telescope or the discovery of penicillin.

Because of Pluto's small size and enormous distance, it is impossible to observe the planet's surface and any possible spots that one might use as reference points in order to determine its period of rotation. Nonetheless, in

1954 and 1955 the astronomers M. F. Walker and R. H. Hardie noticed small periodical fluctuations in the brightness of the planet. This phenomenon can easily be attributed to the fact that some zones on the surface of the planet reflect more of the sun's light and others less, and that, as the planet turns on its axis, these zones alternate, producing the observed fluctuations in brightness. From the period of the light fluctuations it appears that the planet rotates in 6 d, 9 h, 20 min.

Its small size notwithstanding, Pluto must have succeeded in retaining some gaseous envelope. Almost certainly, however, all or most of its atmosphere must have frozen rapidly, because of the planet's very low temperature, due to its distance from the sun: −230°C. In this connection Kuiper noted that Pluto's color is almost identical to that of the sun. In other words, the light that Pluto sends into space is qualitatively the same as the light it receives from the sun; its surface, in reflecting the light, does not change color. This is not true of Mars, for example, which gives off a reddish light very different from the light it receives from the sun because of the red-orange color of its surface. Saying that the surface of Pluto reflects the sun's light without altering its quality is equivalent to asserting that it is white; thus it is almost certain that the planet's surface is covered with immense stretches of ice and snow.

A visit to this planet would be depressing. At night a black sky strewn with bright stars, scarcely scintillating for lack of an atmosphere, looms over endless stretches and huge mounds of ice. Then, at a certain moment, a star much brighter than the others rises, and the landscape lights up much as the earth does in the moonlight: the sun has risen—a sun that should provide light and heat but instead, revealing with its eerie light the desolate stretches of ice surrounding us, increases the gloom that prevails in this distant land.

We have arrived at the bounds of the planetary system. We have come so far from the sun that it appears as a small point (Fig. 2.20), and yet we are still so far from the stars it is almost as if we had never left the earth.

We have explored the planets and made an important discovery. Contrary to what was believed from Galileo's time up until recently, the planets are not like the earth, nor are they similar to one another. Certainly, however, the difference between the planets and the earth is not so great as the ancients prior to Copernicus and Galileo thought. The earth itself is a

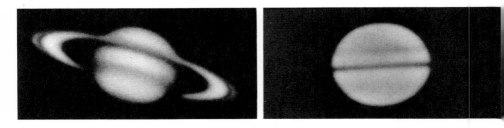

Fig. 2.18 Saturn's rings observed at different angles between 1934 and 1940. This phenomenon is due to variations in the position of the earth. That the rings disappear when the earth is situated in their plane indicates how thin they are; their thickness probably does not exceed 20 km. (*Lowell Observatory*)

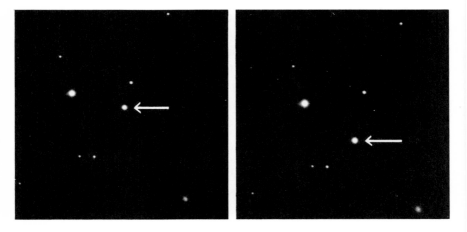

Fig. 2.19 The planet Pluto, a luminous starlike point, revealed by its displacement among the fixed stars over 24 h. (*Mount Wilson and Palomar Observatories*)

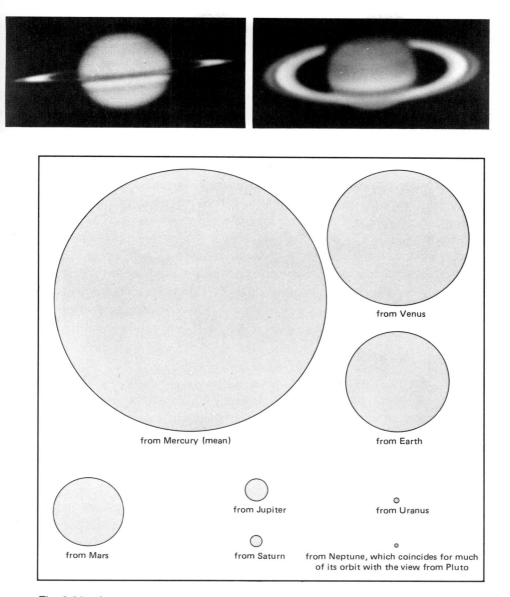

Fig. 2.20   Apparent size of the sun as seen from various planets of the system. The view from Pluto has been omitted, as it coincides, over a large part of its orbit, with that of Neptune.

planet, and certain phenomena—such as revolution around the sun, rotation, the presence (with a few exceptions) of an atmosphere—are characteristics common to all. But when we examine the planets at close quarters in physical terms, we find that among Mercury, Venus, Earth, Mars, and Jupiter there are no two planets alike. Saturn, Uranus, and Neptune appear more or less similar to Jupiter and to one another, but we do not know them as well as we do the planets closer to the sun. The discovery, afforded by the space probes, of the great variety in the appearances of the inner planets, leads to the conjecture that if there are such marked differences between planets of comparable dimensions circulating at comparable close distances around the same sun, there might well be greater differences between distant planets circulating around other stars similar to the sun.

Soon we shall leave the solar system and proceed further into cosmic space, exploring increasingly remote regions of the universe. But before we begin this new adventure, let us round out our knowledge of the solar system, which, as I have said, does not consist solely of the sun and the major planets.

THE SATELLITES

The planets that we have visited are generally not alone. Many of them constitute, in turn, the centers around which other bodies—satellites—revolve. One of these satellites, the moon, has been a focus of man's attention since prehistoric times, yet the modern concept of satellites did not originate until the beginning of the seventeenth century, when Galileo, using his telescope, discovered four bodies revolving around Jupiter.

Today, in the entire solar system, thirty-three satellites are known, and many of them are proving to be as interesting as the planets. In fact physically they are quite similar to the planets. Like planets, satellites are bodies that in general do not radiate light and heat of their own; they may have an atmosphere and constitute, like the earth, an environment that might support life. There is only one characteristic that distinguishes planets from satellites; whereas the former revolve around the sun, the latter have a planet as the center of their motion.

The satellites known today are distributed thus: Earth, 1; Mars, 2; Jupiter,

13; Saturn, 10; Uranus, 5; and Neptune, 2. Mercury, Venus, and Pluto lack satellites—at least, any large enough to be observed with our telescopes.

Satellites are commonly thought to be small heavenly bodies. This is not always true. A satellite is certainly smaller than the planet around which it moves, but since the planets of the solar system are not all equal, it may be bigger than some other planet. In fact Ganymede and Callisto, satellites of Jupiter, are larger than Mercury (Ganymede is almost as big as Mars), and Titan, one of Saturn's satellites, is little smaller than Mercury. The moon is exceeded by these three satellites, by Io and Europa, satellites of Jupiter, and by Triton, a satellite of Neptune (Fig. 2.21). On the other hand, the two satellites of Mars, Phobos and Deimos, are truly miniature moons. Photographs taken from *Mariner 9* in late November 1971 showed that these satellites are not spherical bodies, but rocks, littered with craters and so irregular that one could see the sky even while looking downward, clinging to some overhang as well defined as a balcony (Fig. 2.22). Phobos, the larger of the two, is markedly ellipsoidal in shape, with axes 19, 21.5, and 27 km long. Deimos, smaller and more rounded, has an average diameter of 13 km. One might circle one of these bodies on foot with a few jumps, inasmuch as the lower force of gravity would make a man, on their surface, as light as a cork. But who would have the courage to jump, faced with the prospect of shooting off into space?

An accurate account of all that is known about the satellites would fill a book. However, we would find out only the history of their discovery, information about their orbits, descriptions of some of their physical features (such as their dimensions and apparent brightness), and nothing more, because as yet we have only fragmentary information about their most interesting features—the surface, the atmosphere, whether they might sustain some form of life or constitute an environment in which man could live. In this sense the satellites are almost totally unexplored worlds. Once they are better known, many of them will surely prove much more interesting than the stony desert that is our moon. Ganymede and Titan definitely have an atmosphere; some of Saturn's satellites appear to be covered with ice. Thus these satellites seem to be much more like the earth—and for us, perhaps, more hospitable than the planets around which they revolve.

We are already able to conceive fantastic visions of the world of the satellites. How rich the nights on Jupiter and Saturn must seem, lit up by

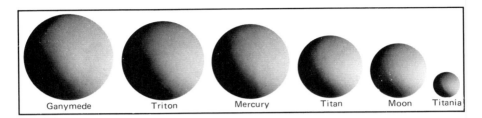

Fig. 2.21   The largest satellites of each planet compared with Mercury.

Fig. 2.22   A striking view of Phobos, the larger of the satellites of Mars, in a photograph taken from the space probe *Mariner 9* in December 1971. Arid and riddled
with craters like the moon, Phobos also shows dreadful precipices from which one
could look either up at the sky or down at the planet Mars, which looms nearby.
(*NASA*)

several large moons, as well as other lesser ones; fainter because they are smaller or more distant, coursing rapidly across the sky! Given what we have learned about the nature of Jupiter and Saturn, we must resign ourselves to the fact that such a spectacle will never be observed by human eyes, just as no living being will ever contemplate the solar system from the surface of the sun. But we might see a vision even more impressive: the view from a satellite of Jupiter or Saturn. From one of those worlds one could see not only the other satellites, lighting up the sky in various ways as they alternate in the sky and their phases change, but also the principal planet around which the satellite is revolving. At certain times this planet would be completely illuminated, immense, full of configurations of various colors; at other times, it would be reduced in partial phases, down to a slender but enormous scythe enclosing a black disk entirely bereft of stars, a kind of circular plaque on the sky, which the planet covers and obscures in the zone on which it is projected. This spectacle, which our imagination may suggest but can only incompletely visualize, may someday actually be admired by man.

THE COMETS

Our view of the solar system is not yet complete. Other heavenly bodies besides the planets, asteroids, and satellites traverse the space around the sun, attracted by the enormous central body that holds them in its power. These are the comets—strange-looking bodies that many people have not seen, yet everyone has heard of (Fig. 2.23).

The first peculiarity to strike the student of comets is the shape of the orbits that they follow. As we have seen, all the major planets, as well as most of the minor planets, move around the sun in nearly circular orbits. Comets do not. Their orbits are highly elliptical, so that, in the course of a single revolution around the sun, a comet may find itself at a minimum distance from the sun considerably less than that of the earth and at a maximum distance greater than that of the most distant planets (Fig. 2.24). The celebrated Halley's comet, for example, comes within 90 million kilometers of the sun and then continues out beyond the orbit of Neptune. Obviously, such a journey takes a long time (over 76 years), and only

Fig. 2.23   The great comet Ikeya-Seki, photographed by W. Liller above Los Angeles on 29 October 1965. Such conspicuous comets are rather rare. Because of the long exposure, we can see the nocturnal glare of Los Angeles, which disturbs astronomical observations.

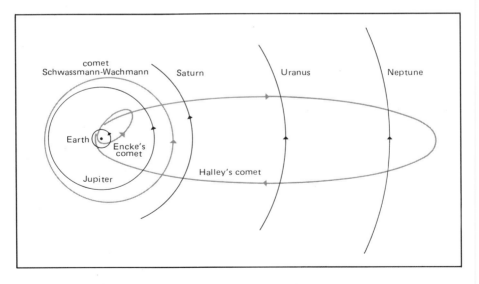

Fig. 2.24   Orbits of the comets Halley, Encke, and Schwassmann-Wachmann projected on the plane of the earth's orbit (the ecliptic). Halley's comet moves retrogradely with respect to the planets.

during the few months when it is near the sun and the earth does the comet become visible to the naked eye.

Some comets have such extended orbits that a single revolution takes hundreds or thousands of years; the comet of 1680, for example, could have been observed previously only 7,000 years B.C., before the start of the great civilizations, and will not be visible again until about 10,500 A.D.

For these reasons the appearances of comets seem altogether sporadic and irregular, and they were believed to be so until E. Halley, at the beginning of the eighteenth century, claimed that the comets that had appeared in 1531, 1607, and 1682 were one and the same. Halley was so sure of his discovery that he predicted that the same comet would appear again early in 1759. Dutifully keeping its appointment, the comet reappeared, to shine on the tomb of the astronomer, who, though he had died seventeen years earlier, had bequeathed his name to the comet forever (Fig. 2.25).

Of course, there are also a great many comets that have much less extensive orbits and reappear much more often, such as Encke's comet, which reaches its minimum distance from the sun every 3 years and 4 months. All of these, however, are much fainter and usually visible only with a telescope.

A number of comets are observed each year, from a minimum of one to a maximum of seventeen.[1] Some have never been observed previously; others are on return visits. The fact that comets, while belonging to the solar system, have very long orbits accounts for the irregularity of their appearances but does not explain their unusual formation or why they are so different from the planets. However, this will be adequately explained when we interpret the numerous physical observations that have been made so far.

According to the findings from these observations, and according to current opinion, comets consist of a brighter, more compact central portion, the nucleus, surrounded by a lighter, somewhat elongated envelope, the coma, which trails off in a third zone, the tail, composed of highly rarefied gases which disperse into space. The first extraterrestrial observations, carried out

[1] The record years for comets have been 1970 and 1975, during each of which seventeen comets were observed.

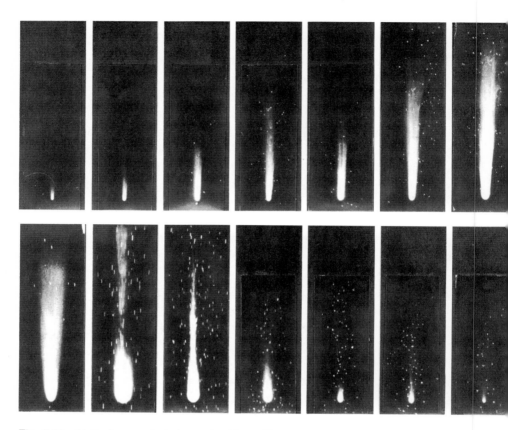

Fig. 2.25   Halley's comet photographed from Mount Wilson Observatory during its most recent approach. Note the evolution of the coma and tail, as well as the variation in the speed of the comet, shown by the varying length of the star trails during the various exposures.

by means of artificial satellites, have shown that the head of a comet can be enormously more extensive than the coma. Ultraviolet observations of Comet Tago-Sato-Kosaka, made from space on 14 January 1970, showed a brilliant envelope of hydrogen extending over 800,000 km from the nucleus; thus the head of this comet, which when observed by ordinary means seemed to have a diameter of 150,000 km, was in reality bigger than the sun. Since that time, large envelopes of hydrogen have been found in all the comets that could be observed from vehicles outside of the earth's atmosphere. The nucleus, with a diameter of about 10 km, is a collection of solid bodies, primarily metallic particles and frozen substances (Fig. 2.26). This iceberg-like structure is porous; vacant spaces occupy up to 95% of the total volume. Despite the enormous dimensions that the coma and the tail can attain—the exceptional comet of 1843, including the visible part of the tail, reached a length of 320 million km—the total mass of a comet is usually very small, often much smaller than that of our planet. The average density is also extremely low—less than that of a vacuum produced in our laboratories. Thus, the definition that the astronomer J. Babinet gave comets, *"des riens visibles,"* is fully justified. Less reasonable is man's recurrent fear concerning the possibility of an eventual collision. Unless our planet is hit by a nucleus (an extremely improbable event, although it has already occurred at least once, in 1908), mankind will never undergo the slightest inconvenience, since passage through the tail of a comet would result, at most, in a spectacular rain of shooting stars.

Comets do not normally present the familiar aspect of a body with a tail. At great distances from the sun, in the remote depths of space, a comet is reduced to an enormous sphere composed of elements that become progressively denser toward the center. When comets approach the center of the solar system, repulsive forces coming from the sun act on the lighter portions of the envelope, displacing the strata that are less dense with respect to the nucleus and thus forming the coma and tail. Therefore, the coma and tail are always situated away from the sun: when the comet is approaching us, the head is leading; when it moves away, the tail is in front (Fig. 2.27).

In short, it is the sun that clothes the comet, and the aspect under which these bodies appear to us depends mainly on solar activity (in addition to

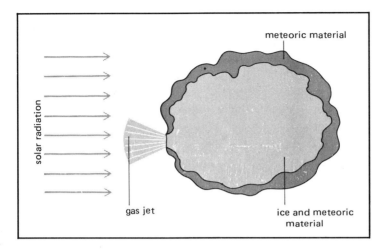

Fig. 2.26   Outline of the nucleus of a periodic comet according to Whipple's model; it would be composed mainly of dirty ice covered by a layer of meteoric material.

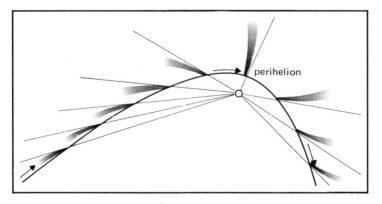

Fig. 2.27   Orientation of a comet's tail, shown in various positions along the comet's orbit; it is always directed away from the sun.

other causes inherent in the comet itself); that aspect generally varies from one evening to the next and will even change over the course of a few hours (Fig. 2.28).

Some comets, in their accelerating rush toward the sun, succeed in penetrating its outer layers and in defying the flames of the chromosphere. Eight such comets are known to date. Five were already being studied in the last century; the others were observed in 1945, 1963, and 1965. The most recent of these is Comet Ikeya-Seki, which the Japanese astronomer Moriyama was able to catch at two moments of its impressive solar adventure (Fig. 2.29). According to B. G. Marsden, who has compared the orbit of this comet with that of comet 1882 II, both derived from a comet, observed in 1106, that split in two during one of its passages near the sun.

From what we have seen so far, the comet is not just a fleeting apparition that is embellished with a splendid coma as it rushes at a dizzying speed toward the sun. The story is not complete, for the comet does not fall onto the sun but turns around it and then flees from it into the distance, like a stone hurled from a catapult. This sequence recurs tens or hundreds of times, at intervals of a few years or several millennia, and each time that the sun clothes the comet with a coma and tail, it makes use of matter drawn from the comet itself: this matter is blown away and scattered in interplanetary space. At each return the comet shrinks, and eventually, when it has lost a major portion of its original mass, the forces of disintegration prevail over gravity (which tends to keep the comet together), and the matter the comet contains is dispersed along its orbit in the form of stones, pebbles, gas, and dust, which continue to move in space in an enormous swarm. When our planet passes near one of these swarms in the course of its revolution around the sun, many particles are pulled in by the force of gravity and fall toward the earth. When they enter our atmosphere, they burn up and dissolve, leaving a momentary bright trail in our sky. Thus do some comets spend their last moments, appearing to us as falling stars or meteors (Fig. 2.30).

Here ends our excursion into the solar system. Starting with the body nearest the earth, the moon, we then encountered the largest body, the sun; we continued our journey to many smaller bodies (planets, asteroids, satellites, and the light, evanescent comets), only to discover finally that,

Fig. 2.28   Comet Mrkos (1957d) photographed with the 60-cm reflector of the astronomical station of Loiano (Bologna) on the nights of 21, 26, and 27 August, respectively. Comet Mrkos had two tails: one curved like a scimitar (presumably because of dust), which maintains the same appearance for a long time; the other long and straight, gaseous and blue in appearance, and rapidly evolving.

Fig. 2.29   Comet Ikeya-Seki, photographed on 21 October 1965 from the Japanese astronomical station of Morikura, while it was grazing the sun, passing by it at a distance of almost 464,000 km. The photographs were taken at 02:20 and 03:27 universal time, respectively. That the comet was moving at very high speed can be seen by comparing the two photographs, taken a little over an hour apart. The tail lagged behind considerably in relation to the conjunction of sun and comet, for the speed of the head of the comet was much greater than that with which the matter issuing from the nucleus moved away to form the tail. (*Courtesy of F. Moriyama*)

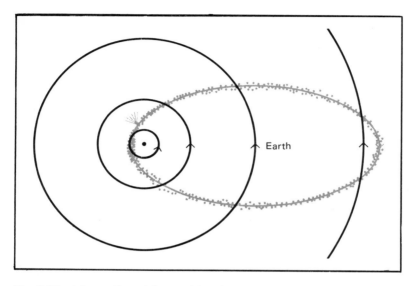

Fig. 2.30   Intersection of the earth's orbit and that of a comet: when the earth crosses the point of intersection, a meteor shower occurs; this shower is usually given the name of the comet.

together with all these objects, large and small, there are very small particles—pebbles and dust—which revolve around the sun and are part of this enormous system, some 12,000 million km in diameter.

Given the fact that at this point the mind begins to have difficulty picturing the actual dimensions and distances, let us draw a new picture of the solar system in space, reducing all the dimensions to one ten-billionth of their real value.

In this reduced-scale model the sun becomes a ball of 14 cm diameter, and the earth is represented by a seed of sesame 15 m away. Alongside the earth, 4 cm away, is a tiny sphere of 0.3 mm diameter—the moon. At 78 m from the sun we shall place Jupiter (a marble will do), and at 600 m Pluto, a grain of millet smaller than that representing the Earth, indicates the limits of the planetary system. The world of the stars lies far beyond these limits. Before meeting the nearest star (another sun—or, on our scale, another ball), we should have to travel 4,000 km. And that is only to meet the first star. The others would be represented by spheres approximately the same size as the preceding ones, but much more distant—so distant that within a radius of 200,000 km we should hardly count one hundred.

The significance of this outcome is truly striking; indeed, it begins to suggest to us the idea of the "cosmic void," which is so often referred to incorrectly, as it was when the first astronauts ventured out beyond the earth's atmosphere. In the picture reduced to one ten-billionth in scale, in which the world on which we were born and where we have spent our entire existence becomes hardly a seed of sesame, and the enormous solar system, whose limits are still inaccessible to exploration by map, is a zone hardly 600 m in radius, the stars, reduced to small bodies some 10–20 cm in diameter, are so rare that we would count barely one hundred in a sphere with a radius equal to half the distance from the earth to the moon. If we try to imagine such an enormous sphere and picture placing in it, here and there, only a hundred objects smaller than a football, we can get some idea of the immensity of cosmic space—that space which contains the world of the stars, far beyond the limits of the solar system.

But that is not all. We have seen that at the distance of Uranus the sun was reduced to a star, albeit a very bright one. At the limits of the solar system the sun's presence in the sky is not sufficient to dispel the darkness of the night, and in order to find another source of light we must approach the nearest star, traversing a distance eight thousand times longer than the already enormous distance that we have traveled from the sun to the edge of the solar system. And we must traverse this same distance again, if we wish to proceed from that star to another. We shall have scarcely begun our journey when we shall find ourselves in darkness, a darkness that is not gloomy and somber but dotted with myriad stars, in a darkness like our blackest nights.

Night—this, then, is the normal aspect of cosmic space. Day, during which we lead most of our active lives and which is so normal to us that it seems impossible that there are places where it never exists, is not a general phenomenon but a very rare event that can occur only on the surface of a planet that is near its sun and has an atmosphere that diffuses the sun's light. Elsewhere, everywhere, night reigns.

Night, but not darkness; for like the starry night that follows the day (when one sun disappears to show us thousands more), this cosmic night—so like the earthly one—opens up to us the pathways of the firmament, leading toward new earths and new suns, toward mysterious, impressive, and splendid objects, toward the bounds of the universe, where we shall eagerly try to discover the alpha and the omega of all creation and, by discovering the meaning of the whole, understand also the significance of our own existence.

Leaving the solar system behind, we shall begin by moving to the nearest star, Alpha Centauri. Traveling with the average speed of the spacecraft that now go to the moon and back, the voyage would take 500,000 years. Traveling at the speed of light, at which speed it would take a little over one second to reach the moon, the voyage would take 4.3 years. If we simply let ourselves go in imagination, we could be in that region of space in a few instants—the time it takes to shut the windows, in order to cut out the street noises, and to turn off the light in the room. The less there is to remind us of the earth, the stronger the suggestion will be, and the more vivid the impression. Since we cannot approach the fiery surface of the star, let us land on one of its planets, which we shall assume is like the earth and similarly located in relation to its sun. This is just the start of our journey, and it is better not to find ourselves abruptly stranded in surroundings too strange.

We arrive at night, and find that the sky appears fairly familiar, which helps us to feel at home. Naturally, we do not see the moon or the planets of our solar system. Perhaps we see instead some other planet revolving around Alpha Centauri, like the planet we are on. A few stars appear slightly fainter or brighter than they appear from the earth and have shifted slightly in the sky, but on the whole we recognize all the constellations that we are accustomed to seeing from our window at home, and the belt of the Milky Way still traverses the sky amid the same stars that mark it out in the terrestrial sky. We have traveled one hundred million times the distance that separates the earth and moon, a journey that would have lasted half a million years at the speeds now reached in space travel, yet in the world of the stars our motion represents a scarcely perceptible step; the appearance of the heavens remains much the same. If we look in the direction of the constellation of Cassiopeia, however, right where it adjoins the neighboring constellation of Perseus, we see a yellowish star, about as bright as Rigel or Procyon, that we never noted in our sky. This star is the sun. From this distance it still looks impressive, and while it adorns the nights of the planet to which we have come as one of the brighter stars in the sky, it also shows us the part of the universe from which we set out and in which the earth, with its cargo of life, can be found.

But time is passing. As we contemplate the sky, the night becomes less dark, and at a certain point a glimmer of light appears along the horizon and steadily grows brighter. It is the dawn—a dawn quite like ours, lighting up the landscape, painting the sky, and finally culminating in the triumphant light of sunrise. We are amazed; the risen globe seems to be our sun. Has that body which just a moment ago seemed reduced to a star suddenly returned? The light that illuminates the landscape has the same color and intensity as the light of our terrestrial day, and the luminous body slowly rising in the sky has the same appearance and even the same diameter as the sun. But it is not the sun; it is Alpha Centauri, the star nearest the earth and almost identical to the sun in terms of appearance, mass, and size. Naturally, the likeness is complete, since the planet on which we find ourselves moves in an orbit similar, even in its dimensions, to that which our earth follows around the sun. The surroundings now seem less familiar than they did at night, because we can see the landscape, which is quite different from that of Earth. However, the light that illuminates the landscape and the heat that warms it are the same as those we were accustomed to on earth, and it occurs to us that the universe is not, after all, so extraordinarily varied as science fiction has described it.

This judgment is too hasty. As the sun, which seems so familiar, approaches the midpoint of its daily journey, we are struck by an unexpected sight. There on the horizon appears a second sun, risen we know not how or when. It is a little larger than the first, but much fainter; thus its dawn was lost in the light of the first sun and its rising passed unnoticed. It emits an orange light verging on red. Now everything casts two shadows, and the colors of objects vary according to whether they are in the shadow of the first sun, in that of the second, in both shadows, or in regions illuminated by both suns. It is impossible even to accustom oneself to this fantastic landscape, for as the two suns move across the sky, the effect of one, then the other, prevails; thus the landscape and even we ourselves are continually changing shadows and colors. Given the variety of shades and colors that earthly objects assume when illuminated by a single sun over the course of a day, we can easily grasp what must happen during the slow permutations of two suns. As the afternoon of the first sun advances, the second sun begins to predominate, and when, after it has passed the meridian, it alone remains, everything appears immersed in a new, pale reddish light, and objects cast only one shadow. After a few hours the second sun also sets, rapidly, in a sea of fire, making way for the familiar night sky. After a few hours of night the show begins again, and a new day with two suns returns to inspire our wonder. This cycle goes on for several days, but not forever. The picture we have drawn is not permanent, because the planet we are on completes its revolution around the first sun in a year, and the relative positions are constantly changing. At this point the first sun may set during the afternoon of the second, but after the planet has traversed about one-quarter of its orbit, it will be opposite the second sun, and the two suns will rise, culminate, and set together; at that time we will be able to witness the remarkable sight of one sun being eclipsed by part of the other. After about three months the orange sun will rise first, and the yellow one similar to ours will follow. Finally, after another three months, when the planet is aligned between the two suns, each will rise as the other sets, and there will be no night; rather, a yellow day and an orange day will alternate continually until, little by little, the suns will return to the positions they were in upon our arrival (Fig. 3.1). Calendar makers on this planet will have to be very skillful—all the more so since the description just given of the motions of these bodies is not yet complete. Indeed, the two suns we have watched "moving" across the sky (as a result of the planet's motion) are in fact not fixed in space but move, in turn, around the system's center of gravity, completing one turn every 80 years. During this period they approach within 11.2 A.U.[1] of each other and are separated by a maximum distance of 35.5 A.U.

We stated that our planet revolves around the brighter component; thus the brightness and apparent diameter of this sun remain constant. The orange sun, however, will at times appear brighter and larger, or fainter and smaller, and, of course, in addition to the "year" in which the two suns move back to their original positions, we will also observe a "year" of 80 years, during which time the orange sun, besides changing in color and apparent size, will slowly cross the entire sky.

DOUBLE STARS

Before continuing to explore the system of Alpha Centauri, we would do well to pause for a moment, to place in context what we have seen and to

---

[1] The mean distance of the earth from the sun is called the astronomical unit (A.U.): 1 A.U. = 149,600,000 km.

recall how we came to know about this double sun, even if it means exploring and studying other systems of this kind. The argument is somewhat
complicated, but we shall succeed in finding satisfactory answers to nearly
all our questions before this first stop among the stars is over.

A double star like Alpha Centauri is far from rare. Many stars that appear
single to the naked eye appear as double stars when seen through a
telescope (even a modest one). The first double star was discovered on 7
January 1617 by Benedetto Castelli, a friend of Galileo; informed of it,
Galileo observed it in turn a week later. This star is Mizar, the central star of
the three in the handle of the Plough. Interestingly, this star appears double
even to the naked eye: those endowed with good eyesight can spot a
companion, much fainter and very close, north of the brighter star. However, it is not this double star that interests us here; this closeness is merely
a matter of perspective. We are concerned, rather, with the brighter star,
which, if we were to observe it with even a small telescope (as Castelli and
Galileo were the first to do) would divide into two stars; further observations
would show them to be connected with each other.

Various other double stars were subsequently discovered, but it was not
until 1779 that the great astronomer W. Herschel started to observe them
systematically. Herschel believed that the stars appeared in pairs only
because of perspective and, assuming that the fainter member of each pair
was the farthest from us, he set out to determine possible displacements of
the brighter star (attributable either to its actual motion in space or to the
parallax effect). With this data he could then deduce the distance between
the stars.[2] A comparison of measurements made between 1782 and 1804
convinced him, however, that many double stars could not be the result of
mere perspective juxtapositioning of stars actually very distant from each
other. If that were the case, when positions measured some years apart
were compared, one would observe either no displacement at all or else a
rectilinear one. Instead, in many cases Herschel found that one of the stars
had moved around the other along the arc of an ellipse—exactly what an
outside observer would observe while measuring the position of the earth

[2] The methods of determining distances between stars are described briefly in
appendix A. This information, however, is not necessary for an understanding of
what follows.

relative to the sun over the course of a year. Thus, as early as 1803, Herschel was able to announce that many of the double stars he had observed were single physical systems; he was even able to give the periods of revolution for five of them.

Following up on this great discovery, W. Struve began the golden age of the study of double stars in 1822. Through a visual check of 120,000 stars (all those down to apparent magnitude 8.5 visible above his horizon), accomplished with the 23-cm refractor of the Dorpat Observatory, Struve was able to discover and measure over three thousand double stars. He continued this research at the Pulkovo Observatory, to which he was appointed in 1839, and it was later carried on by his son Otto. It would take too long to describe all the subsequent research in this field. We will simply mention that S. W. Burnham, A. Hall, E. Dembrowski, G. Schiaparelli, E. S. Holden, R. T. A. Innes, and, more recently, R. G. Aitken, W. J. Hussey, W. H. van den Bos, and R. Jonckheere, among others, discovered an ever-increasing number of double stars, extended the observations to the Southern Hemisphere, and carried out an enormous number of positional measurements.

The observation of double stars is so long and laborious that the task often requires the work of several generations. In fact, once a double star has been discovered and the position of one component relative to the other has been measured, the same measurements must be carried out repeatedly at given intervals. By comparing the various positions, one can then determine whether one of the two stars has moved, and, if so, whether it has moved along a straight line or part of an ellipse (Fig. 3.2). Obviously, only in the latter case can one be sure that the double star is a physical system. Unfortunately, two conflicting limitations make it very difficult to arrive at this point. From Kepler's third law we know that when one heavenly body moves around another (for example, when a planet moves around the sun or one component of a double star moves around the other), the farther away it is, the more slowly it moves along its orbit. Thus if a star is revolving near its companion, its motion will be rapid; as the distance increases, however, the ellipse in which it moves becomes longer, and the star moves more and more slowly. As a result, in certain cases it may take a star tens, hundreds, or even thousands of years to complete a revolution, and in order to discover any physical connection between a pair of stars, it

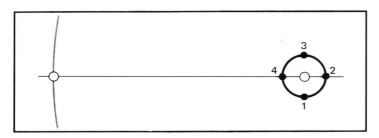

Fig. 3.1 The visibility of two suns from a planet orbiting around the yellow star of the Alpha Centauri system. It is assumed that the planet is similar to the earth and as far from the yellow star as the earth is from the sun and, further, that it moves counterclockwise around its axis and in its orbit, the plane of which coincides with that of the orbit of the orange (shaded) star. When the planet is in position 1, it sees the orange sun rising at the yellow sun's noon; at 2 the two suns rise, culminate, and set together; at 3 the orange sun rises first, and at its midday the yellow sun rises; at 4, finally, each of the two suns rises as the other sets. In this last position the hypothetical inhabitants of the planet experience no nights but rather alternate "reddish" and "yellow" days.

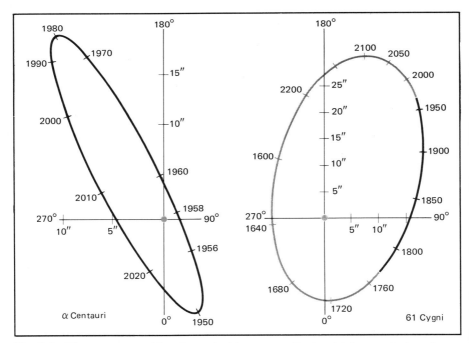

Fig. 3.2 The apparent orbits of two double stars: α Centauri (observed, since it was discovered in 1689, for more than three revolutions) and 61 Cygni, whose orbit has been observed so far only to the extent indicated by the heavier line.

may be necessary to wait and observe these stars over years, decades, or even centuries. Unfortunately, such cases are the rule. Under more favorable conditions (when the stars are closer together and the revolution is more rapid), we are confronted with the other difficulty: the stars' apparent separation is so miniscule that they cannot be distinguished, even with the most powerful telescopes, and often their double nature is not even discovered. The only exception is those double stars (even close ones) which are sufficiently near the earth. It is precisely the study of this reasonably homogeneous sample that will enable us to get an idea of the percentage of double stars relative to the total number of stars.

The task of those astronomers who study double stars is thus to measure, wait, remeasure, and often to leave to later generations the satisfaction of drawing conclusions. Some sow, others reap. The situation may seem discouraging, but in reality it is not. Such collaboration with those who have gone before and with other scientists not yet born extends Man in time, allowing him to be active beyond the limits of his own lifetime; it unites in the same research not only contemporary astronomers of various races or nationalities but also scientists of different epochs. Thus the astronomer who witnessed Napoleon's empire, the scientist who miraculously escaped the destruction of the Pulkovo Observatory in Leningrad in the siege of 1942, and the scientist who will measure the positions of double stars from an observatory set up on the moon may someday sit around the same table, each with files of observations, discussing the ramifications of a problem that absorbed them all equally, a problem they contributed equally toward solving.

Meanwhile, a catalog-index of all double stars observed visually up to 1960 was published a few years ago, in order to present a conspectus of the work done so far on double stars and to indicate those for which observations were most urgently needed. There we find, along with other information, the position of each pair in the sky and its designation in the various catalogs, the number of times it has been observed, the years of the earliest and most recent observations, all the known physical characteristics of its components, its distance from the earth, and the motion of the system in space. Of course some of these data are often lacking since they have not yet been obtained, but even so, the enormous importance of the catalog is obvious: it not only lists everything that has been done, it also indicates what remains to be done.

This catalog shows that the double stars observed visually and measured up to that time numbered 64,247. The number of those for which it has been possible to determine an orbit and to establish with certainty a physical bond is very much smaller—696 as of 1970. The periods obtained are generally of the order of a few decades. So far the shortest is that of the star Wolf 630, 1.7 years (less than the orbital period of Mars); the longest (among the more reliable figures) is that of Eta Cassiopeia, 480 years.

If we know the distance of a double star, its period of revolution, and the length of the major semiaxis of the orbit, we can determine—using Kepler's third law in the generalized form Newton gave it—the sum of the masses. In fact, starting with the universal law of gravitation, we obtain Kepler's third law in the following form:

$$K\frac{a^3}{P^2} = m_A + m_B$$

where K is a constant, a is the major semiaxis expressed in seconds of arc, P is the period, and m_A and m_B are the masses of the two stars. If we were to observe the sun-earth system at the same distance, we would have the analogous relation

$$K\frac{a_E^3}{P_E^2} = m_E + m_S,$$

where the symbols "E" and "S" indicate that the same quantities refer to the earth and the sun. From the two equations we find that

$$\left(\frac{a}{a_E}\right)^3 \frac{P_E^2}{P^2} = \frac{m_A + m_B}{m_E + m_S}.$$

Given the fact that $a_E = \pi$ by the definition of parallax (see appendix A), that the mass of the earth, m_E, is negligible assuming that the mass of the sun is equal to 1, and measuring the period in years (in which case $P_E = 1$), we have the formula

$$\left(\frac{a}{\pi}\right)^3 \frac{1}{P^2} = m_A + m_B,$$

from which we see that, knowing the distance (parallax), the major semiaxis of the system, and the period, we can derive the value of the right-hand side of the equation, which is precisely the sum of the masses expressed in units of the solar mass.

Thus if the positional measurements are carried out so as to determine not the position of the fainter star relative to the brighter star but the position of both stars relative to other stars in the field, then we obtain not the relative orbit of the fainter star but the absolute orbits of both—that is, the orbits of both components about their common center of gravity. In that case, given the fact that the distances of the two components from the center of gravity are inversely proportional to their masses (as we learn from mechanics), we can immediately deduce from the ratio of the major semiaxes the ratio of the masses. By combining these two results (the sum and ratio of the masses) in a system of first-order equations that a high-school student could solve, we are able to determine the individual masses.

We have succeeded, then, in weighing the stars. This is a stupendous feat, possible only in the case of double stars. There is, in fact, no other method for determining the masses of individual stars.

So far we have only been able to weigh some hundred stars—very few compared to the vast number of stars populating the heavens, but they are enough to give us an idea of the situation.

It turns out that, in general, these stars weigh about as much as the sun. Stars ten times as massive as the sun are rare, as are stars one-tenth as massive as the sun. Therefore, those stars having masses some tens of times greater than that of the sun are altogether exceptional.

Here are some extreme cases: The heaviest stars known are the two components of the binary system HD 47129 (Plaskett's star); their masses were recently found to be more than 55 times that of the sun. The lightest stars are the two components of Luyten 726-8 B (much better known as UV

Ceti). According to the observations of the astronomer W. J. Luyten, whose name the pair bears, the mass of each star is scarcely four-hundredths that of the sun. They are indeed extremely light stars, yet each is 13,300 times heavier than the earth.

EXCEPTIONAL STARS

WHITE DWARFS

The most extraordinary discovery made in weighing the stars was the discovery of the companion of Sirius, that brilliant star that shines in the southern sky during winter nights.

As early as 1844 F. W. Bessel, observing the motion of Sirius in space, found that it was not rectilinear but oscillated periodically about a mean position, and he therefore deduced that the motion of the star must be perturbed by another heavenly body of considerable mass (Fig. 3.3). In 1862 the unknown perturbing body was discovered visually and named Sirius B. It is a small, very faint star, which, since it is relatively close to us (less than 9 light years away), could not be intrinsically very luminous. Hence, it was thought that Sirius B must certainly be a star of low temperature, and therefore a red star.[3] We can imagine the astonishment of astronomers when in 1915 W. S. Adams discovered that Sirius B is on the contrary a white star. This fact cast a different light on things. As is well known, if we have two surfaces with the same area, one white and one red, the white one will be much more luminous than the red; thus the white color of Sirius B means that, contrary to what was thought, the whole surface of this star is very luminous; and if it appears so faint, despite this fact and despite its proximity, we must be dealing with a very small star.

Further research showed that Sirius B is, in fact, almost twice as large as the earth but contains so much matter that it weighs as much as the sun. It follows that the star's density is incredibly high, about two hundred thousand times that of water. On this star a piece of matter the size of a matchbox would weigh eight tons. Since so great a mass is concentrated in so small a sphere, the surface gravity is enormous. If a man could be transported to the surface of Sirius B, in order to lift a mass of one gram he

---

[3] The relation between color and temperature will be explained later.

would have to exert the force that would be required to lift a mass of two hundred kilograms on the earth. But at any rate a man could not survive on this star for as much as an instant (even if its surface temperature were as low as that on Earth) for he would be literally crushed by his own weight.

So far over fifteen hundred stars similar to Sirius B have been discovered. They are all very close to us, and there is reason to believe that many more exist that we do not know about, simply because, being small, they are very faint. Indeed, various statistics have shown fairly consistently that there must be a little over two white dwarfs for every hundred normal stars. However, it has been possible to determine the exact mass for only three of these (Sirius B, Procyon B, and o_2 Eridani B). These do not appear to be the densest white dwarfs, though. Among the twenty-five best-known white dwarfs, the one that holds the record for density is LP 321-98, a small star that appears to us a million times fainter than the faintest stars visible to the naked eye; its density is more than three hundred times greater than the enormous density of Sirius B.

Those unfamiliar with the "strange" discoveries of physicists and astronomers will doubtless become somewhat skeptical when they read that a matchbox can contain enough matter to weigh eight tons; most people are convinced that matter could not possibly be so condensed. In fact astrophysicists themselves were astonished, given the fact that on the earth there are at most twenty grams per cubic centimeter, and even on the sun, where matter is quite condensed, there are no more than a hundred grams per cubic centimeter. On the other hand, the implications of the observations were clear, and there were no doubts as to the validity of the discovery. The theoretical interpretation posited by physicists is that the matter inside these stars must be in a particular physical state that can be neither solid, liquid, nor gas. Such a state (now known as "degenerate") is considered to be a fourth state to which matter can be reduced when it finds itself in certain conditions. Let us explain this with an example. Suppose we accumulate an enormous quantity of eggs in a large silo and keep adding more. As long as there are only a few eggs, nothing will happen, but at a certain point those underneath will break, overcome by the weight of those above, and after the breakage, that weight will occupy a smaller volume. This is what happens to the atoms of which matter is composed. In the gaseous state they wander in space, separated by considerable distances.

In bodies in the liquid and solid state these distances diminish, so that the atoms are more or less in contact with one another—like the pile of eggs. But when the pressure reaches extremely high levels (equal to about ten million atmospheres), the atomic structure gives way and collapses, and matter, while neither gaseous nor liquid, is no longer solid either. The properties of this degenerate state have been studied primarily by Enrico Fermi; degenerate matter is also known as "Fermi gas." In these conditions matter no longer radiates energy. Even if raised to extremely high temperatures, to hundreds of millions of degrees, it will appear dark and cold to the outside observer.

Thus the white dwarfs appear dark and cold to us, except for an outer layer in which the matter is not degenerate. This layer, being luminous, has allowed us to see these stars and has permitted us to discover the strange wonders of their interiors, which no one will ever see.

MULTIPLE STARS

Let us return to the world of normal double stars. Here yet another surprise awaits us: there exist not only double stars but also triple, quadruple, and even sextuple stars. These systems appear truly enchanting in the telescope, especially when the components are of different colors. The spectacle they afford the inhabitants of the planets of these systems must be something quite unimaginable, surpassing even the two dawns and two sunsets that we witnessed on the planet of Alpha Centauri!

But if we return to observe the sky from that planet, we shall make another important discovery—Alpha Centauri is also a triple system. We did not realize this fact at first, because from that planet the third sun appears very faint—like the faintest of stars. It is in fact a red star, intrinsically very faint and also very far from the principal pair. It has long been observed from the earth and noted as a very close star. According to some measurements, it seemed even closer than Alpha Centauri; hence, it was named Proxima Centauri. It is now certain that these three stars are located at nearly equal distances from the sun and that they form a single system. According to the astronomer C. Gasteyer, Proxima is at least 6,700 A.U. from the two central stars. That means that it completes one revolution around the other two in not less than 367,000 years.

PLANETS OF THE NEAREST STARS

This is what one star, the one nearest to us, has taught us. We have discovered the world of double stars and have watched unusual sights from a planet remarkable for the mere fact that it belongs to that system. After so many strange experiences, we must find out whether the planet we have invented really exists. Well, we are not sure whether there is a planet near Alpha Centauri, but one thing is certain: other planets do exist, near other stars, outside the solar system. In other words, the planetary system to which our earth belongs is not the only one in the universe.

In fact, ever since it was discovered that the stars are bodies like the sun and appear as small luminous points only because of their enormous distance, it was thought that they could also be centers of planetary systems analogous to our own.

Thus as early as the end of the last century, it was widely believed that the number of planets in the universe similar to the earth might be enormous, even larger than the number of the stars, since a star might have at least five or six planets. Yet no such planet had ever been seen. Actually, they have yet to be seen, and, I hasten to add, it is almost certain that no human eye will ever see them, at least from the earth. To understand why, let us resort again to an example: Suppose—assuming the most favorable circumstances—that Alpha Centauri does indeed possess our hypothetical planet, that it is the same size as Jupiter, and at the same distance from the central star as Jupiter is from the sun. We can then calculate that it would appear to a terrestrial as a star of magnitude 23, barely accessible to the most powerful telescopes on Earth, but at any rate it would be literally obliterated in the field of the telescope by the brilliant light of the principal star.

If things go so badly for the nearest star and a planet as big as Jupiter, we can understand how much worse the situation would be for more distant stars and for planets smaller than Jupiter. Thus there is absolutely no way we can see planets in other systems, at least with the means currently available. So astronomers have tried to get around this difficulty. They have taken the view that it is not really necessary to see these planets; the important thing is to discover whether they exist, which is not the same thing. In fact, the method of perturbations (which led, in the last century, to

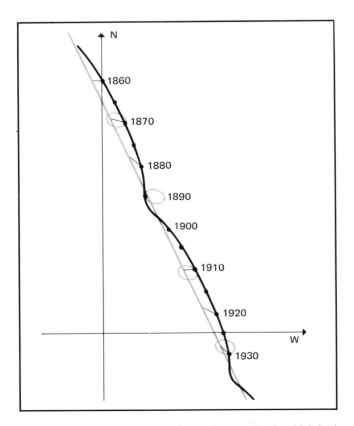

Fig. 3.3   Apparent motion of Sirius (heavier line), which is the resultant of a uniform rectilinear motion combined with an elliptical one (lighter lines).

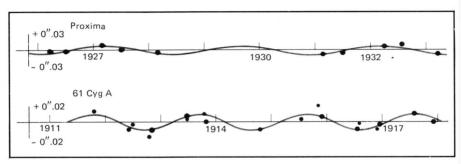

Fig. 3.4   Perturbed motion of two stars, Proxima Centauri and 61 Cygni A: the displacements from a rectilinear motion are probably due to the presence of a dark companion.

the discovery of Neptune and of Sirius B) is still applicable in this case, if conditions are optimal. In this way one can not only discover the existence of an invisible and unknown body, but also calculate its mass and the distance that separates it from the star around which it revolves (Fig. 3.4). In 1937 the Sproul Observatory began a program of observations of all nearby stars, based on such a method. The first success was obtained with the double star 61 Cygni. Near one of the two components a dark body having eight times the mass of Jupiter was discovered. This body moves around its sun in 4 y, 9 mon, 15 d, at a mean distance of 2.4 A.U. (that is, a little less than the distance of Jupiter from the sun). More recently, another possible planet was found near "Barnard's star": by examining over two thousand plates taken at Sproul Observatory between 1916 and 1962, the astronomer P. van de Kamp found that the motion of this star in the sky is not rectilinear but follows the serpentine course typical of perturbed bodies.

These two bodies represent our most complete and certain results. For the moment we do not know much more. But it must be noted that the apparent displacements due to perturbations are very small, and measuring them, even under the favorable conditions presented by the nearest stars, means straining the limits attainable with our instruments. We need only consider, for example, the fact that in the case of the planet of 61 Cygni the maximum displacement is barely 0".02. This corresponds to the angle subtended by an object one meter high at a distance of ten thousand kilometers.

In considering these difficulties, we must keep in mind the fact that the number of stars near which dark bodies have been found is relatively high. In fact, of the fifty-nine stars nearest to us, eight may have planets, and those being observed are the most suitable candidates for observation—that is, they are heavy. Other possible planets having lesser masses escape us, just as the smaller planets of our solar system would be invisible if we tried to observe them from a distance greater than that of the nearest star.

Thus other planetary systems exist, and they certainly outnumber those that our limited means have allowed us to discover so far. Hence the stellar universe is much more thickly populated than it appears to our eyes, and the bodies we do not see are, unlike the stars, precisely those on which life could develop—perhaps even intelligent life. It is like watching city lights from afar in the darkness of night and reflecting that for each light there are so many other houses that shows not even a single light, much less an inhabitant. But we can infer in that brightly lit anthill the existence of a population that rejoices, suffers, works and plays, is active and rests, is born and dies. Thus when we now look at the starry sky, we cannot help feeling that those lights may also illumine settlements and activities unseen by us, and that for other beings those sparks of light, which for us are merely decorative, perhaps represent life.

The existence of other planetary systems is also significant in another sense: it would indicate that if one day we were to succeed in developing our technology to the point where we could leave the solar system, we might have destinations to aim for, landing places for our spaceships, and, perhaps, new earths and new skies for mankind. Maybe even the hypothetical planet of Alpha Centauri exists, and perhaps some day human beings will be able to land on it and contemplate that phantasmagorical sight which a moment ago, guided by science, we saw only in our imagination.

TYPES OF STARS

When we were on the planet of Alpha Centauri, we saw that, while one of the stars of that system is yellow like our sun, another is orange, and a third is red, and that the latter two are fainter than the first. Thus stars do not all have the same color or the same brightness. By examining this question more closely, we shall discover matters of great importance.

Let us begin with color. There are blue, white, yellow, orange, and red stars—along with, of course, all the intermediate shades. Corresponding to these differences of color (apparent even to the naked eye, if we look carefully) is another, more fundamental difference, that of the spectra.

It is well known that if one observes the light of a luminous source (for example, the sun) through a spectroscope, one sees a continuous band with the colors of the rainbow (violet, indigo, blue, green, yellow, orange, red) crossed by numerous dark lines, each of which corresponds to a certain element or compound. This band is called the spectrum. The continuum, with the various colors, originates in the high-temperature source that emits the light; the lines originate in the lower-temperature elements of the gas that lie between us and the luminous source. If there is no gas in front of the incandescent body emitting the continuum, no dark lines are seen; if there is no luminous body and only gas at a certain temperature, one would see only the lines, which in this case would stand out brightly (in "emission") against a dark background.

The spectra offer an effective means of discovering the chemical composition of stars and other heavenly bodies. For example, in order to determine whether there is hydrogen on a star for which we have a spectrum, we need only obtain in the laboratory the element's spectrum, which will show a certain series of lines (for example, one in the red, others in the green, others in the blue, etc.). If we find the same lines in the spectrum of the star, we can be sure that hydrogen is present in its flaming envelope of gas. If we then measure the intensity of the lines, we can even find out how much hydrogen is present. The same process works with any element.

Now, when we observe the spectra of stars of various colors, we note marked differences. The spectra of blue stars show essentially the lines of ionized helium; those of white stars, the hydrogen lines. Yellow stars are rich in metals; red stars in molecular compounds of carbon, in titanium oxide, and in zirconium oxide.

The difference between the colors, as well as between the spectra, is caused by the variation of a very important physical parameter—the temperature. Who among us has not paused for a moment to observe a welder at work? Under the increasing effect of the heat of the torch, fed by the air tanks, the iron to be worked becomes at first red, then yellow, and at last almost white. The stars show the same progression: red stars have relatively low temperatures; yellow ones high; and blue ones very high. There is a physical law ("Wien's law") that applies to the stars as to any other body. This law underlies the different relative intensities of the various colors and hence of the different colors of the stars.

If, however, we wish to explain the fact that certain elements appear in some spectra and different ones in others, the direct interpretation of the mechanism will help us more than the forge analogy.

The gas of which the stellar atmosphere consists is, like all matter perceived by the senses, composed of atoms. As we have already seen, the atom can be pictured as a kind of miniature solar system, made up of a positive nucleus, which contains most of the mass, and a certain number of peripheral electrons having a total negative charge equal to the positive charge of the nucleus; thus the whole is electrically neutral. Atoms can unite to form a molecule (two atoms of hydrogen may form a hydrogen molecule, two atoms of hydrogen and one of oxygen may form a molecule of water, etc.). However, the link binding the atoms together, or the electrons to the nucleus, can be broken. For example, think of a three-atom molecule as a group of three magnetized pellets stuck together. If a bullet is fired at it and hits the group with sufficient energy, it can make one of the pellets break away and may even break up the group altogether: the three pellets would be dispersed enough to escape their mutual magnetic attraction, and the group would not re-form. Something similar happens in the stars.

The temperature of a body is due to the fact that the particles comprising it (whether atoms or molecules) are in movement; the greater the energy supplied to them, the more their velocity—and hence the temperature of the body—increases.

Now, if there are many molecules of titanium oxide in the atmosphere of a star, this means that the temperature is lower than that necessary to bring the particles to a velocity such that collision with a molecule of titanium oxide would break the bond, separating the titanium from the oxygen. If we find neutral metals, it means that the temperature is low enough so as not to allow one or more electrons to be torn away from the nucleus. If an electron is removed, however, the element is said to be ionized; it can be ionized one, two, or more times, depending on whether it has lost one, two, or more electrons. Thus, if we know, from laboratory experiments and theoretical physics, the energy of dissociation of the molecules and the ionization potentials of the various elements present in the stars, we can deduce the temperature from their presence or absence.

Moreover, the difference in the intensities of the lines of the various elements in different types of stars does not indicate an actual difference of composition between one star and another.[4] Rather, it simply indicates a difference of temperature, and therefore the temperature can be determined with more precision than is possible when color alone is used.

The different spectral types have been arranged in various classes,

[4] There may be such differences, but the effect is a more recondite one and their discovery more laborious.

named according to the letters of the alphabet and accompanied by a number (from 0 to 9), so as to permit a more detailed classification within each class. In order of decreasing temperature the classes are

O   B   A   F   G   K   M

(English-speaking astronomers use the mnemonic "Oh, be a fine girl, kiss me.")

The principal results obtained from the preceding considerations are exemplified in Fig. 3.5. The intensity of the lines varies; those of hydrogen are strongest in the spectrum of Sirius (type A0); from type K0 on there is considerable formation and development of molecular bands.

Let us now consider the subject of brightness. For greater precision we shall henceforth not say merely that some stars are bright and others faint; we shall indicate their relative brightness numerically. The scale that we shall use was introduced two thousand years ago by Greek astronomers and, with a few small amendments and extensions, is still in use today. In it the value 1 is given to the brighter stars and the value 6 to the faintest stars visible to the naked eye; intermediate values are given to those in between. These values are called apparent magnitudes. The addition of the adjective "apparent" indicates that they merely give us an idea of how the stars appear as seen from the earth. In fact, stars may differ from each other in magnitude either by differences in their intrinsic brightness or because their distances from us are different; we must be able to measure either intrinsic brightness or distance to be able to determine both quantities.

The number that expresses the magnitude of a star does not tell us directly how much fainter one star is than another, because this is in fact an ocular sensation, and, according to a law formulated by G. T. Fechner, we know that when a sensation varies in arithmetic progression (1, 2, 3, 4 ...), the corresponding stimulus varies in geometric progression. In other words, stars of magnitude 6 are not five but one hundred times fainter than those of magnitude 1, and so on, by a factor of 100 every time the magnitudes differ by five units.

Another strange characteristic of this scale is that, because of the way it is constructed, the fainter a star is, the higher the number expressing its brightness. Thus the faintest stars that can be observed with the largest telescope in the world are of magnitude 23, while the brightest stars visible

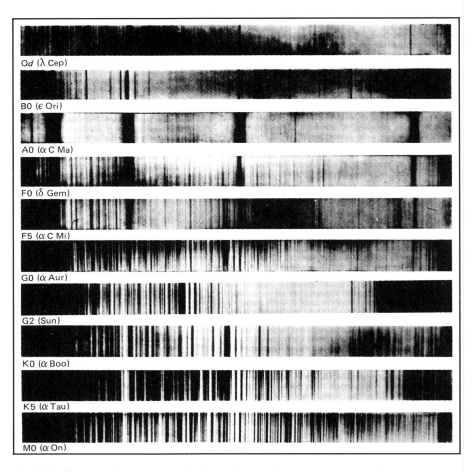

Fig. 3.5   Spectra of some stars (indicated in brackets), according to the Harvard classification. Note the progressive increase in the number of lines. (*University of Michigan*)

to the naked eye, which are 4,000 million times brighter, are of magnitude 0. The sun, moreover, is 43,000 million times brighter than the brightest stars.

This is all fairly obvious. We know that the sun is by far the brightest star, and to say that it is 10^{20} times brighter than the faintest star that can be seen with the most powerful telescope does not really tell us anything new. It would be quite another matter if we could determine exactly how the sun's intrinsic brightness compares with that of other stars. In other words, if we could determine the apparent magnitude that each star (including the sun) would have if it were at a specific, uniform distance from us, then a comparison of the magnitudes would tell us directly to what extent a given star is really brighter or fainter than another. In order to do this, astronomers have introduced the so-called absolute magnitude. This is the apparent magnitude that a star would have if it were 32.6 light years away from us. Clearly, from what has been said, in order to determine the absolute magnitude we must know the apparent magnitude and the distance. When these data are known, the absolute magnitude can be obtained by means of a simple formula. (Those interested in further information may refer to appendix A, where it is also explained why the distance chosen may appear somewhat strange.) But this is merely a technical matter. It is more interesting to actually compare the luminosities of the stars that we know—including, first of all, the sun.

At a distance of 32.6 light years the sun's apparent magnitude would be 4.8; it would still be visible to the naked eye, but very faint. Many stars that are normally seen with the naked eye would be invisible. Other stars, however, even some that are now invisible, would become dazzling, a hundred or a thousand times brighter. If all the stars visible to the naked eye, while maintaining the same positions in the sky, were moved nearer or farther from us so as to be all at a distance of 32.6 light years, the sky would be unrecognizable. Some stars, such as Sirius and Altair, would be only slightly fainter; others would disappear; many, on the other hand, would appear very much brighter. The constellations would look entirely different and the sky would become splendidly bright, like an amusement park at night. More than five hundred stars would appear brighter than Sirius, now the brightest star in our sky, and some would be fully a hundred times as bright. If we were to see such a spectacle at night, we should at first be taken aback, but then we would try to find out the reason for such a change, which in fact would bring us much closer to reality than what we ordinarily see, falsified as it is by the different distances of the stars. Clearly, stars can be brighter or less bright than the sun, or they may differ considerably in size (from the sun or among themselves), or, finally, both cases may occur.

The clearest and most complete answer is supplied by a famous diagram discovered independently by the astronomers E. Hertzsprung and H. N. Russell, whose names it bears. The Hertzsprung-Russell (H-R) diagram has proved to be one of the greatest astrophysical discoveries of this century and one of the most effective means of investigating the mysteries of the stellar universe, the nature of the stars, and their past and future (Fig. 3.6).

In order to understand the diagram better, let us construct our own. Let us represent, by a point on a plane, each star—or rather the two characteristics it has that we have learned, namely the spectral type and the absolute magnitude. The first will correspond to the horizontal axis, the second to the vertical axis. Contrary to what we might expect, the points do not lie scattered at random over the entire plane but are distributed mainly in certain zones, forming a sort of inverted 7. Let us consider first the branch descending from left to right, called the main sequence. Bearing in mind the fact that the absolute magnitude indicates the intrinsic brightness of a star and the spectral type its temperature, we notice immediately that type O stars are hotter and brighter than those of type M; the same holds true, to a lesser degree, for the other spectral types. That is quite natural and could easily have been predicted, for everyone knows that the surface of a body at a certain temperature emits much more energy (particularly light) than that of a body at a considerably lower temperature. But let us now try to explain the group of stars above to the right. Here the preceding interpretation does not work. We find, in fact, type M stars—that is, stars that at a relatively low temperature (about 3,500°K) shine as brightly as type B stars, which are instead at high temperatures (about 20,000°K and even greater). There can only be one explanation: given that the temperatures of both faint and brighter M stars must be the same, the quantity of light emitted per unit of the surface (e.g., per square meter) should be identical in both cases. If, then, one of the stars is more luminous, its total surface must be greater. In other words, the more luminous M stars have larger diameters than fainter

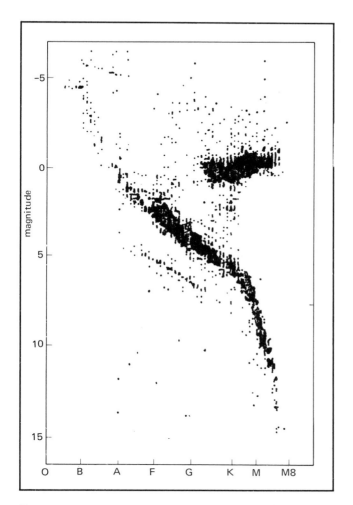

Fig. 3.6 Hertzsprung-Russell diagram (from the *Astronomicheskie Zhurnal*), with the spectral type as abscissa and the absolute magnitude as ordinate. Most of the stars are arranged in a zone inclined obliquely from left on top to right below, called the main sequence. There are two other groups on opposite sides of the main sequence, corresponding to the giant stars (*above*) and the white dwarfs (*below*).

ones. Calculations made on this basis show that the more luminous M and K stars must be much larger than other spectral types. Astronomers have named the larger group giants and the smaller ones dwarfs.

Thus stars can have a wide variety of diameters (Fig. 3.7). This is true not only of the red stars among the two groups but also for those of the main sequence. In fact, B and A stars, even if they do not attain the size of the red giants, have been found to be larger than those of more "advanced" spectral types (that is, moving from type O to type M along the main se-quence, the stars are progressively smaller). This effect therefore reinforces the effect of the difference in temperature. Thus dwarf M stars are fainter than B or A stars, not only because they are less hot but also because they are smaller.

The diameters of stars can also be measured directly; methods vary depending on the circumstances and unfortunately are not universally ap-plicable. Direct measurements have confirmed the results obtained with the H-R diagram. The following quantitative results were obtained.

First of all, it was found that many stars have approximately the same diameter as the sun. In general these stars belong to the main sequence of types F and G, those of type F having slightly larger diameters and those of the last subclasses of type G having slightly smaller ones. Type A stars have diameters about 2 times that of the sun; B stars about 8; and type O stars about 12. M stars, on the other hand, have diameters only half that of the sun. The stars of the giant branch are all greater in diameter than those of the same type in the main sequence. The diameters of F giants are about 4 times that of the sun; G giants some 10 times; and M giants up to 40 times. Stars have also been found with diameters greater than those of the giants, and they are called, naturally, supergiants. Their diameters range from 75 times that of the sun, for types O and B, to 35 for type F, and up to 280 times for type M. Stars in this last group are truly enormous. If one of them were placed at the center of the solar system, it would extend beyond the earth's orbit, overrunning the earth, Venus, and Mercury. There are other extreme cases. The smallest stars are the white dwarfs, with diamet-ers little greater than that of the earth. The largest of all are some double stars, such as ε Aurigae and VV Cephei, with diameters over 1,000 times the diameter of the sun. VV Cephei, which is the largest known star, has a diameter 1,900 times greater than the sun's.[5] If placed at the center of our planetary system, it would overrun nearly all of it, reaching as far as Saturn. Only the outermost planets—Uranus, Neptune, and Pluto—could warm themselves in the red light from this enormous sun, which would nearly cover their sky.

If we are to form a complete picture of the variety of stars that we have encountered so far, there is one other small point we must consider. We have seen that the masses of the stars are never more than some ten times the mass of the sun, and we have just learned that their volumes, on the other hand, can be thousands of times greater than the volume of the sun. The obvious implication is that the densities of many stars, especially those of the giants and supergiants, are extremely low. The differences in density between one star and another are thus greater than those between the masses and the volumes. We find at the lower extreme stars with densities hardly two- or three-hundredths that of the sun and at the other end the white dwarfs, a hundred thousand to ten million times denser than the sun.

THE NEIGHBOURHOOD OF THE SUN

This, then, is what we have discovered about the stars from an initial exploration of the solar neighborhood. To obtain a clearer, more concise picture, let us see how many stars and what kind of stars we encounter in the part of space that is closest to us, in which there are stars that are relatively faint and even some objects that are not directly visible. We shall consider as our neighbors all the stars not more than seventeen light years away. Within a sphere of this radius, with the earth at its center, we shall find forty-four stars in all.

To get a better idea of the distances, let us make a scale model, with 1 cm representing 1,000,000 km. Starting at the earth we find at a distance of 2.5 mm the moon, and at 1.5 m the sun. At 7.78 m we meet Jupiter and at 59 m Pluto, whose orbit denotes the confines of the solar system. Now space becomes emptier, and we have to travel fully 410 km before we reach Alpha Centauri, another sun and perhaps the center of another

[5] According to measurements made by B. F. Peery (1966), its diameter is 1,620 times that of the Sun.

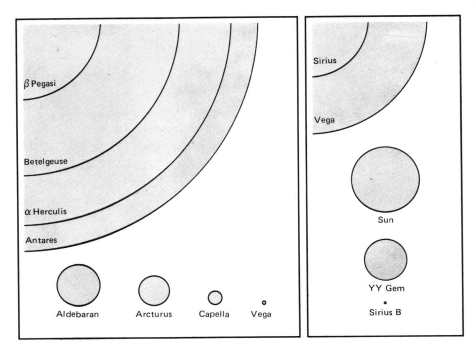

Fig. 3.7 Diameters of certain stars compared with one another, including that of the sun, on two different scales; Vega appears in both diagrams.

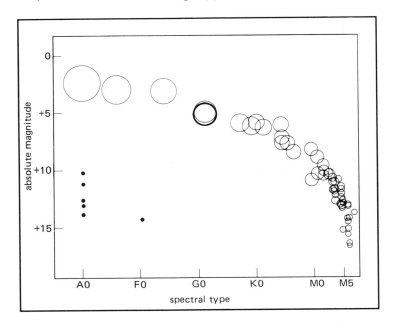

Fig. 3.8 The Hertzsprung-Russell diagram for the nearby stars. The circles indicate the approximate relative diameters: the black circle corresponds to the sun; the other, almost superimposed, to α Centauri A. The points to the left represent the white dwarfs (not to scale).

planetary system. Heading in another direction, we come across Barnard's star at a distance of 570 km, and in yet another direction, at 820 km, we find Sirius, the brightest star seen from the earth. Scattered here and there we find various other stars—61 Cygni at 1050 km, Procyon at 1070 km, Altair at 1590 km—until, at a distance of 1620 km, we encounter the star Stein 2051, which marks for the moment the limit of our journey.

Such are the proportions of the distances between the earth and the nearby stars, those which Man is now hoping not only to know better in the next few years, but even to visit. The distance to be covered is indeed great. The road we have traveled so far from here toward the nearest star corresponds to a distance of 2.5 mm on the road from Rome to Genoa, and in relation to the farthest of the nearby stars it is as if we had covered the first 2.5 mm on the road from Rome to Copenhagen. Yet even in the region of the nearer stars, the limitless spaces of the cosmos do not end but are just beginning. Clearly, those who call our present astronautical attempts "cosmic voyages" do not know the meaning of the word *cosmos.* No one would claim to have made transoceanic voyages after wading in the surf. This comparison does not mean we should disparage recent accomplishments, which are sublime indeed in view of human dimensions; rather, it is intended to help us adjust to a new scale in spatial dimensions.

But let us return to the stars in our neighborhood. Of the forty-five stars present, including the sun, thirty-two are single, eleven are double, and two triple. The total number of stars in this sphere having a radius of seventeen light years is thus sixty. Of those, at least seven (besides the sun) have small, planetary-type companions. To obtain a better physical picture, let us construct the H-R diagram for all of these stars, and, in order to visualize the situation better, let us represent each star as a circle having a diameter proportional to that of the star (Fig. 3.8). We notice first of all that the sun (black circle) is one of the largest and brightest stars. Apart from Alpha Centauri, which is practically a twin to the sun, only three stars exceed the sun in size and brightness: Sirius, Altair, and Procyon. Six of the nearby stars are white dwarfs, and very many others are red dwarfs. Completely absent, however, are red giant stars and supergiants, red or blue. If we find no giant stars in a randomly selected volume of space (and in fact have no reason to suppose that our neighborhood is a region with special characteristics), this simply means that giant stars are very rare. And if, in spite of this fact, we see so many in the sky, it is simply because they are so bright that they are visible even at great distances. When, therefore, continuing our exploration of the universe, we find myriad stars, for the most part very distant and of high luminosity, we must keep in mind the fact that, on the basis of the proportions we have found for the nearby stars (not to mention the hundreds and thousands of giant stars that we can see or photograph), there must be millions or thousands of millions of red dwarfs that we cannot see, because of their faintness and enormous distance.

Our first contact with the world of the stars is now completed. The stars that we have met appear to us in a great variety of colors, masses, diameters, and densities, but nearly all of them can be considered normal stars. One might imagine that, in order to get an idea of the population of space, it would suffice simply to multiply the number of these objects, but this is not the case. Stars of the types we have learned to know in the neighborhood of the sun exist also in other parts of space, but in addition to these there are abnormal stars, there are groups of stars, and also other objects, small or large, dark or luminous, that are not stars. To become acquainted with them, we need only leave the immediate neighborhood of the solar system and venture out ever farther. We shall thus continue our journey and ultimately encounter even the rarest of these strange objects.

ECLIPSES OF STARS

HIGHLY INCLINED ORBITS

Up until now, in speaking of the orbits of double stars, we have always pictured them the way they would appear, reduced in scale, in a drawing on a piece of paper viewed from above. Orbits do not generally appear this way to us; to do so, however, the plane of an orbit must lie perpendicular to our line of sight (that is, a straight line drawn from our eyes—or the earth, which amounts to the same thing—to the primary star). Obviously we would come across such an orbit only by chance. Ordinarily we observe orbits inclined every possible way. For some the situation is exactly the opposite of the first case; that is, the line of sight lies in the very plane of the orbit, and the secondary star does not seem to follow an elliptical course around the primary star but seems to oscillate back and forth along a straight line.

What we have said does not present an insurmountable obstacle to the calculation of the true orbit, for there exist mathematical methods that enable us to derive it from the observed orbit, which is distorted by the effect of perspective. The subject does, however, lead us to an interesting consideration, which follows from the extreme case when the orbit is viewed exactly in profile.

In this case it is evident that during each revolution the secondary star passes alternately in front of the primary star and behind it, so that from the earth we see two stellar eclipses. The first eclipse occurs when the secondary star eclipses the primary star, and vice versa for the second eclipse. If we measure the total brightness of the two stars when they are separated and also during the eclipses, we shall find that in the latter measurement the brightness is diminished. If both components are visible, this experiment is self-evident.

Let us suppose now that the two components are so close together that they cannot be separated even in the largest telescopes. In that case, the binary property can be discovered from the existence of the eclipses. From the way in which the phenomenon develops and from its duration, one can draw other conclusions as well—for instance, what kind of eclipses they are, which star is the brighter, and even the relative diameters of the stars. To obtain these results, it is necessary to use particular methods and carry out certain kinds of measurements, which we shall outline in only a general way.

The first step is to construct what is known as the light curve. When, after repeated observations of a star, one discovers that its light is variable, the light must be measured systematically and a graph must be constructed in which one coordinate represents the times of observation and the other the observed magnitudes. The resulting graph, called the light curve, shows us how the light of the star varies with time, just as a clinical graph shows the fluctuations of a patient's temperature on different days and at various hours. If the variations consist of regular reductions of the light, repeated systematically at fixed intervals, we can be certain we are dealing with the eclipses of two bodies, and the star is called an eclipsing variable, or, more pompously, a photometric binary. The form of the light curve can be one of the two types shown in Figure 4.1, in which, above each significant section of the curve, are shown the corresponding positions of the stars, relative to the earth, in the course of their orbital motion.

On the right Figure 4.1 shows a total eclipse alternating with an annular eclipse, and on the left it shows what happens when the orbital plane of the two stars is slightly inclined to the line of sight from the earth, in such a way that the centers of the two stars are never aligned with the line of sight and the eclipses are only partial.

Naturally, whether the eclipses are partial or total depends not only on the inclination of the orbit to the line of sight but also on the radii of the two stars. Clearly, if the eclipse is central, it will always be alternately total and annular, whatever the ratio of the radii. On the other hand, the two eclipses can still be total and annular if the radius of one of the stars is sufficiently small compared to that of the other (Fig. 4.1, *right*), but if the radii are more or less equal, or if even at the maximum phase of the eclipse the centers of the stars remain far apart, the eclipses will be partial (Fig. 4.1, *left*).

In both figures the light curve shows two minima: one is deeper, corresponding to the passage of the fainter star in front of the brighter star, and the other less deep, corresponding to the disappearance of the fainter star. These two minima are called primary and secondary, respectively. The minima depend not only on the brightness of the two stars but also on their relative dimensions. For example, the effect is similar if a large, not very bright star is occulted by a smaller, much fainter star, or if a small, not very bright star is occulted by one much bigger but of low surface brightness. Astronomers are able to recognize and separate these two effects, as well as other secondary phenomena. Though they must take these effects into account if they wish to obtain complete, precise results, we can neglect them without forfeiting an overall picture of the phenomenon.[1]

In some cases an effect has been observed that reveals a strange fact: the light curve shows no secondary minimum. This indicates that the less bright component is very faint, or even dark. There exist, then, such things as black stars! However, such an assertion cannot be made in exactly these terms, for such bodies would not come under the definition of stars. Let us simply call them dark bodies, as big as stars and physically linked to them.

Such are the discoveries to which the eclipsing double stars have led us. The visual double stars had opened up new territory for our imagination; now the variety is even greater. The combinations that we can envision are nearly inexhaustible, and the panoramas that we can imagine seeing in the vicinity of each of these systems are innumerable: pairs of blue or yellow stars, or red giants; pairs of assorted stars of various dimensions and colors; small blue stars accompanied by enormous dark red spheres verging on black, which disclose their presence only by intercepting the light of their brilliant companions, or by reflecting it weakly—not stars but ghosts of stars.

In nearly all these pairs the components are extremely close to one another and their revolution around their common center of gravity reaches dizzying speeds: the periods of revolution are generally only a few days, often just a fraction of a day.

These systems, so different from our planetary system (simple and poor as it is, revolving round a single star of a very common type, neither very large nor very small), force us to recognize that nature's variety quite surpasses our imagination; even when guided by science, we always imagine something more modest than what reality subsequently reveals to us.

---

[1] Among these effects we mention the limb-darkening effect due to the fact that the stars—as we saw in the case of the sun—are surrounded by an atmosphere at a lower temperature than the underlying strata, which causes a darkening of the peripheral zones; the phase effect, in which the fainter star reflects part of the light it receives from the brighter one; possible tidal effects, when the stars pass each other at their minimum separation; and so on.

VERY CLOSE PAIRS

The dizzying motions of these bodies are neither inexplicable nor irregular. They are simply the result of their great proximity to one another. As we have already seen, Kepler's third law establishes that the revolution of one heavenly body around another is more rapid when the two bodies are close to each other, and inasmuch as these stars are so close together that they are not individually distinguishable even in the largest telescopes, their motions are very rapid. That being the case, we can make another interesting observation, this one spectroscopic.

According to a well-known effect, called the "Doppler effect," if a source emits a sound of a certain frequency when it is approaching an observer, the latter will hear a sound higher in pitch than the sound he would hear from the same source at rest; when the source is moving away, the pitch is lower. A similar phenomenon occurs in optics. If a source of light is moving toward the observer, all the lines in the spectrum, corresponding to the various frequencies, appear at a higher frequency (that is, they are displaced toward the violet), whereas if the source is moving away, the same spectrum appears shifted toward the red.

Eclipsing binary stars are systems whose orbital plane contains the line of sight from the earth; thus sometimes one of the stars will be moving away from the earth, and, sometimes it will be approaching. When spectroscopic observations are made in different phases of the period, the lines are seen to move alternatively right and left with respect to a mean position. If, then, both components are bright enough to show a spectrum, we will see a composite spectrum (due to the sum of the two) in which, with the passage of time, one group of lines oscillates in one direction and another group in the opposite direction (Figs. 4.2a, 4.2b).

Thus here we have another powerful means of investigation for discovering close double stars. By means of eclipses we can in fact discover only those binaries for which the angle of inclination of the orbital plane to the line of sight is nearly zero. As this angle increases or diminishes, the centers of the stars no longer coincide exactly, and when the angle, while remaining small, still varies a little, the two stars, instead of eclipsing each other, pass slightly above and below each other. In such a situation there is no eclipse, and the binary property cannot be determined on the basis of a change in light; the orbital plane, however, is close enough to the line of sight to produce a strong radial component (that is, in the direction of the earth) in the velocities of the two stars: the effect of the periodic oscillation of the spectroscopic lines may be quite conspicuous, and the binary nature of the star may thus be revealed. Such a star is called a spectroscopic binary.

To sum up: We find that there are two types of binary besides the visual type (those that can be distinguished in the telescope): eclipsing and spectroscopic binaries. The difference between the latter two types and the first type consists essentially in the different method of observation, due to the fact that the stars of the eclipsing and spectroscopic binaries are very close together and the planes of their orbits happen to be only very slightly inclined to the line of sight from the earth.

Because of the particular situation in which these stars are observed, their orbital elements cannot be determined completely as in the case of visual doubles. On the other hand, it is possible to obtain interesting new data about the stars that make up the systems. Thus in the case of photometric binaries, it is possible not only to derive the period, eccentricity, and inclination of the orbit, but also to determine the radii of the two components, using the distance between the stars as unity. In the case of spectroscopic binaries, it is possible to obtain, in addition to the period and other orbital elements, a certain quantity $a \sin i$, which includes the distance a between the two stars and the angle i made by the orbit with the plane normal to the line of sight from the earth. This is not a great deal, but it serves to give us an idea of the distance, all the better since it is expressed directly in kilometers. We obtain besides, still combined with i, two quantities proportional to the masses of the two stars: $m_1 \sin^3 i$ and $m_2 \sin^3 i$.

The ideal situation would be to be able to combine photometric and spectroscopic observations. In fact, if a binary has both characteristics, with the first method we can measure the radii of the two components, taking the distance between the two centers as unity; most important, we get the angle i. Then with the second method, if we can see both spectra, we get $a \sin i$, and since i is now known, we can immediately obtain the value of a in kilometers. Finally, since a was in fact the unit we used to measure the radii of the stars, now that we know the value of a in kilometers, we can also determine the radii of the stars in kilometers. By an analogous process, knowing i, we can obtain the two masses.

Thus we have managed to discover that a given star is double, trace the

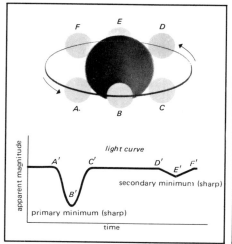
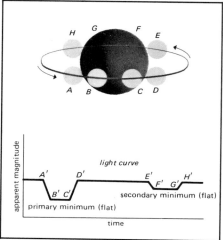

Fig. 4.1   Eclipsing binaries: *left,* partial; *right,* annular. *Below,* the observed light curves corresponding to the two types of eclipse. In each case the smaller star has a lower surface brightness.

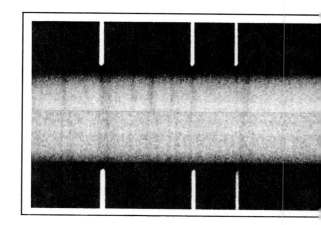

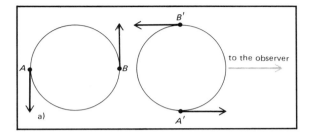

Fig. 4.2A   Two stars of equal mass, in rapid motion about their common center of gravity, observed in the plane of their orbit. At some instants (positions A, B) they appear to move only transversely, and the components of their velocity along the line of sight of the observer are zero; at other instants (positions A', B') one appears to move toward, the other away from, the observer.

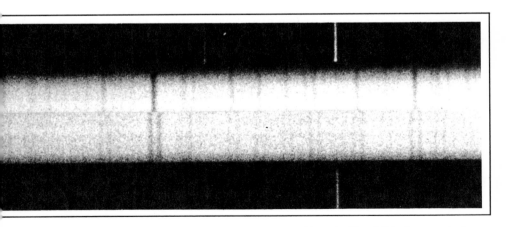

Fig. 4.2B   Corresponding to the two extreme positions in Fig. 4.2A, the spectral lines of the stars (recorded in a single spectrum) appear single, because they are superimposed, or double, as shown in the two spectra of Mizar obtained at an interval of 2 d. The period of this double star is 20.5 d; the separation of the two lines in this figure corresponds to a difference in orbital velocity of 140 km/sec. (*Mount Wilson and Palomar Observatories*)

orbit that the components travel, measure their radii and the distance between them in kilometers, and, finally, weigh the two stars. We have accomplished all this without even knowing the distance of the system from us and without even having seen the stars as two disks with perceptible diameters—all we saw was a single luminous point whose apparent diameter was quite inappreciable. We have to admit it is rather amazing.

Unfortunately, double stars of this type (known as spectrophotometric) are rare. So far scarcely 100 are known. Spectroscopic binaries with calculated orbits, on the other hand, number 737; eclipsing binaries 4,062. (These latter two values were taken from catalogs published in 1967 and 1969, respectively.) The number of eclipsing binaries is so much higher simply because it is easier to discover and observe them; in fact they can be distinguished rather easily by comparing photographs of a given region of the sky taken at suitable intervals.

Such photographs enable us to compare thousands of stars, even very faint ones, simultaneously. In order to discover spectroscopic variations, however, we must compare the various spectra of each star, obtained one at a time, and this is only possible with the brighter stars. Actually, if we think about it, spectroscopic binaries ought to outnumber photometric binaries, since the latter correspond to that fraction of the former that show eclipses because of the particular value of the angle *i*.

COMPLEX SYSTEMS

We said a while back that these binaries are double stars characterized by the different ways in which we perceive them. This is not to suggest that the difference between them and visual double stars consists solely in the method of observation. In fact of the two conditions responsible for the eclipse and the spectroscopic variation, one (the stars' particular orientation with respect to the earth) has no bearing on binary comparisons, but the other (the proximity of the two components and their high velocity of revolution) entails physical consequences of utmost importance. In many cases the two stars are so close together that they exert tidal effects on each other. Sometimes matter is actually drawn away from one star and falls onto its companion; in other cases matter escapes from both stars and forms an enormous envelope that surrounds the whole system.

Naturally spectroscopy allows us to discover many of these phenomena, to follow their development, and to form, case by case, an overall picture of what takes place in these distant systems. Each of these explorations is a novel experience and often it will lead to the discovery of a system different from any yet known. A very beautiful system is that of ζ Aurigae, consisting of a red supergiant that periodically eclipses an extremely bright but much smaller blue star. Because of the vast expanse of the red supergiant's atmosphere, prior to the actual eclipse the light of the brighter star fluctuates at length behind the irregular, but increasingly dense, veils of the supergiant's atmosphere. The phenomenon unfolds in reverse at the end of the eclipse (Fig. 4.3).

Another interesting system is that of RW Tauri, a giant orange star around which revolves a much smaller blue star, encircled along its equatorial plane by a rapidly rotating red ring of hydrogen (Fig. 4.4).

No less interesting, if only by virtue of its complexity, is the system UX Monocerotis, which must be a spectacular sight when viewed from a nearby planet. Here one star is white, the other yellow like the sun, but both are much larger than the sun. They are so close to each other that the space between them is full of hydrogen, which emits an intense red light. One hemisphere of the yellow star continuously gives off matter, especially calcium and hydrogen, which then falls onto the white star; the opposite hemisphere of the white star gives off other matter, which ends up on the yellow star. Some of the matter constantly streaming from both stars is eventually dispersed in space and forms an envelope that completely surrounds both stars.

There are numerous other cases: VV Cephei, a red star nineteen hundred times larger than the sun, accompanied by a blue star a hundred times smaller; ε Aurigae, two giant stars, one blue, the other red; Algol, the first double star of this type discovered, which we now know is composed of three stars—two semidetached stars, which whirl around one another in less than 3 days, and a third, more distant star, which completes a revolution every 22½ months.

But what is the use of going on? Each case has a history of its own, and though thousands of cases have already been revealed, millions more have yet to be discovered—an endless host of worlds waiting to become the object of our exploration and our imagination, though perhaps they have already served that purpose for the eyes and senses of other beings.

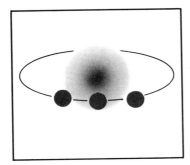 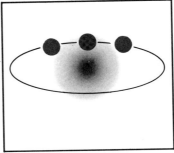

Fig. 4.3   Eclipse of ζ Aurigae: a very bright, relatively small blue star is periodically hidden behind a red supergiant. The outer strata of the latter are so rarefied that they are semitransparent with respect to the light of the blue star they are occulting. The figure (which is obviously schematic) does not take account of the irregularities of the outer strata of the supergiant's atmosphere, which should be pictured as tenuous veils of stellar matter, gradually dissolving in space.

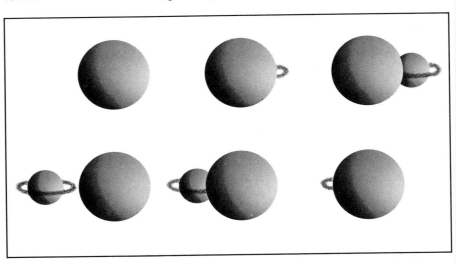

Fig. 4.4   RW Tauri, another strange binary system that would be very interesting to observe close up. A giant orange star dominates the sky; at a given moment a loop, which later turns out to be a ring, emerges from behind one edge; within the ring is a blue star, which also emerges, little by little, from behind the orange star; eventually both stars appear separate and distinct in the sky. After some time the smaller component passes in front of the other star, causing a phenomenon we can hardly imagine: a blue star inside a red ring, both projected against an orange star that seems to enfold them. (The figure outlines only certain successive phases of the occultation of the blue star, leaving the rest of the phenomenon to the imagination.)

But for now let us consider one more star, an extremely interesting one, not only on account of its characteristics, but also because it is the most famous example of a special group of eclipsing variables. The star in question is β Lyrae, one of the first variable stars ever discovered and even today among those most studied, because of the presence of certain secondary phenomena that remain an enigma. The variability of this star was discovered in 1784 by J. Goodricke, an amateur astronomer who died when he was only twenty-two and who, though a deaf-mute, managed to achieve much more in his brief life than do so many people who are able to use their speech and intellect to a very advanced age. Goodricke was the first to regard the precise, regular light variations of β Lyrae and Algol as eclipses. Many of his contemporaries opposed or underestimated him, but subsequent facts proved him right.

The system of β Lyrae, in particular, proved quite strange. The light curve shows periodic eclipses every 12 days, 22 hours, 22 minutes. Between one eclipse and the next, however, the light does not have the extended maximum one might expect. The variation is continuous: from the principal minimum to the maximum, to the secondary minimum, to the second maximum, and finally to the principal minimum, at which point the cycle begins anew (Fig. 4.5).

These facts were interpreted as follows: The two stars are not spherical but ellipsoidal, and they revolve around their common center of gravity in such a way that the major axes of the two ellipsoids always remain aligned; in other words each star always turns the same hemisphere toward the other. On the basis of this scheme the spectral types and the brightness of the two stars, their radii and masses, and the distance between the two centers were determined (see Table 4.1). But spectroscopy, which had contributed to the discovery of these characteristics, also revealed other, unsuspected phenomena, and once more β Lyrae appeared complex and mysterious. Today, after the research of O. Struve, in particular, we can form the following picture (even though many points still need clearing up): β Lyrae is a system of two ellipsoidal stars in rapid motion around each other; their diameters and masses are considerably greater than those of the sun. Both stars are white (one tends slightly toward blue, the other toward yellow), but their equatorial planes are surrounded by two red rings (predominantly hydrogen) connected by a kind of enormous gaseous

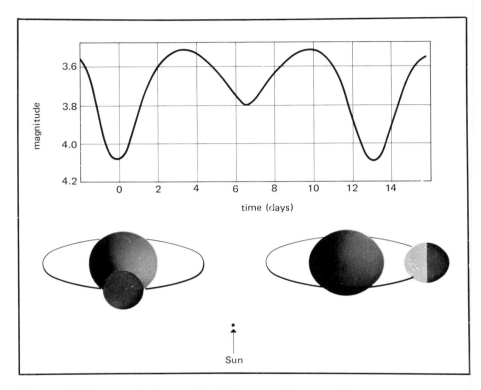

Fig. 4.5   Light curve of β Lyrae. The light curve is interpreted as being due to two ellipsoidal stars that revolve around each other, each keeping the same hemisphere turned toward the other. We are observing the system from a position near its orbital plane. The sun is drawn to scale.

Table 4.1 Characteristics of the two components of β Lyrae

Star	Spectral Type	Absolute magnitude	Radius	Mass
A	B 8	−4.01	19.5	9.74
B	F 5	−1.08	13.78	19.78

According to data obtained by S. Gaposchkin (Handbuch der Physik, L, 231). Assuming the radius and mass of the sun to be equal to one.

bridge—in fact by two enormous bridges side by side, composed of the matter that flows from each star to the other. Both stars are immersed in an enormous, expanding envelope. Indeed, rather than an envelope, it is perhaps more of an immense red spiral, which, in steadily wider but fainter bands, unwinds in the space around the two stars for hundreds of millions of kilometers out to distances of the order of the diameter of our solar system (Fig. 4.6).

Such is β Lyrae, and such, perhaps, are hundreds of other similar stars—no longer spheres, but incandescent bodies rendered oval by the powerful influence of the gravitational forces, bodies from which matter flees, falling from one star onto the other or dispersing in space. Such a sight must certainly be striking. But to those eyes that are perhaps viewing it from a distant planet, it must seem perfectly natural and normal, just as the beauty of a city so often eludes those who live there. Unfortunately, in this case, the tourists who would be able to appreciate these marvels would have to make inconceivable journeys. If a terrestrial spaceship were to set forth for β Lyrae today at the speed of light, it would not arrive there for 1,000 years, and if it were possible to install in the craft a telecamera capable of sending us pictures of these marvels, they would reach the earth not after a second or two, like those coming from the moon, but after another 1,000 years, for 1,000 light-years is precisely the distance from the earth to β Lyrae.

PULSATING STARS

Our excursion in space led us, just now, to a very strange multiple star at a distance of 1,000 light-years from the solar system, a star in which the periodic eclipses of two bodies in rapid revolution cause continual fluctuations in the overall light. We have seen that these fluctuations are only apparent (that is, they are due to the fact that we view these stars from a point near the plane of their orbit), and we have also noted that there are a great many other stars that show fluctuations of light for this same reason.

Without having to travel so far, we can find other stars whose light varies in the absence of any eclipse. Naturally these may also be double stars,but many are single stars like our sun, and the fluctuations occur because their intrinsic brightness actually varies—in some cases abruptly, in others periodically and continually—for unknown reasons and by means of mechanisms only partially interpreted so far.

PULSATING RED STARS

The first star of this type to be discovered was in the constellation Cetus. Its variations in brightness aroused such wonder (this was 1596, and as yet no variable stars, even eclipsing ones, were known) that it was called Mira, which means "Wonderful." Four centuries later we can only agree that this name still fits it to perfection, both on account of the strange things our observations have revealed and because of the complexity of the problems still to be solved.

Mira Ceti is 163 light-years distant from us. It is an enormous body, with a diameter of 556 million kilometers: if it replaced the Sun at the center of our solar system, it would fill space out beyond the orbit of Mars, and our own planet would end up incinerated in its interior. Let us imagine, therefore, that we are on a sufficiently distant planet—at the distance of Jupiter or Saturn from the sun, say—and let us observe it.

We see an immense dark red disk, dominating a sky that is also red, but darker, almost black. The temperature of this disk is extremely low for a star, barely 1,900° K—so low that the outermost regions fade away in clouds composed of molecular compounds, such as titanium oxide and even, in great abundance, water vapor. There exist, then, not only flaming stars whose extremely high temperatures pulverize matter more and more minutely, to the point of breaking up molecules and splitting atoms, but also stars cool enough actually to contain water vapor. This sensational discov-

ery was made about Mira only in 1963 when, with the development of
astronomical observations in extraterrestrial space, it was observed for the
first time from outside our atmosphere. Until then, since such observations
were made by astronomical observatories on Earth, the water vapor present
in our atmosphere concealed that in the stars and prevented its discovery.

Let us now observe Mira from nearby. The temperature is low, the brightness hardly half that of the sun. But as time goes by, things change. Day
by day the star slowly diminishes in volume, like a collapsing balloon;
the temperature increases to a maximum of 2,700° K; the dark clouds of
titanium oxide and water vapor wandering through the star's atmosphere
dissolve; and the luminous globe diminishes in size a little. Its brilliant red
light, however, increases in intensity until it becomes 80 times as bright as
the sun. If there are any planets in its vicinity with life on their surfaces, what
changes that life must undergo in the course of this kind of stellar season,
and what great capacity for adaptation living organisms must possess!

At this point the temperature starts to drop, the diameter increases, the
clouds re-form, and the brightness slowly diminishes, until everything is as it
was in the beginning. The entire cycle described lasts 331 days and is
continually repeated, although not always identically (Fig. 4.7). From the
earth the star is seen to vary between magnitudes 9.3 and 3.5, that is, it
passes from the rank of a faint telescopic star to a bright star visible to the
naked eye. A few months later, the eye will be unable to discern anything at
the point in the sky where the star was once easily visible (sometimes, at
maximum, it is brighter than the polestar). The first observers were right to
call it Mira.

And we are also right to go on calling it by that name. Observing the
spectrum in around 1920, the American astronomer A. H. Joy suspected
the existence of a companion. After learning this, Aitken, a specialist in
double stars, discovered the companion with the great Lick refractor on the
night of 19 October 1923. It was a small blue star of magnitude 10, so faint
as to become observable only when its larger companion was at minimum.
Various astronomers observed it repeatedly during those favorable periods,
but sometimes, though they searched carefully, they could not find it. The
reason was soon discovered: this second star is also variable, fluctuating
not slowly and regularly like Mira, but, it seems, in sudden, violent bursts,

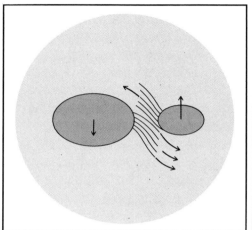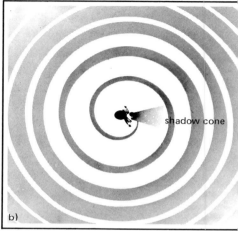

Fig. 4.6   Model of the β Lyrae system. *Left,* the inner part, consisting of the two stars and matter streaming across the two flanking bridges; *right,* on a greatly reduced scale, the enormous red spiral of hydrogen that becomes broader but fainter, until it disappears.

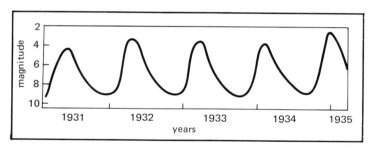

Fig. 4.7   Light curve of Mira Ceti, a long-period, pulsating variable star.

and returning more slowly to minimum, like certain other stars we shall encounter shortly.

Although this second star has been followed now for about half a century, very little is known as yet about its orbit. Certainly the orbit must be quite extended since the star takes such a long time to traverse it completely. In 1961 J. D. Fernie and A. A. Booker, discussing the observations then available, calculated three types of possible orbits, of which, for physical reasons, only the third appears acceptable. In this case the period of revolution would amount to 260 years, and the total mass of the two stars would be 3.9 times that of the sun. The physical characteristics of the blue star indicate that its mass must be at least 2.6 times that of the sun, as has been observed in the case of other stars with similar physical characteristics. Thus the mass of the giant Mira would be of the order of the solar mass. When we consider the immensity of its volume, we arrive at a surprising conclusion: Mira is a body of extremely low density, 200,000 times more rarefied than the earth's atmosphere. Of course the density must be greater in the innermost layers and less in the peripheral ones. Perhaps the nucleus is very small, and we see only the highly rarefied outer envelope, so that we calculate volume and density while including a practically empty intermediate zone. One thing is certain, at any rate: this celestial body, whose distance makes it appear to us as a luminous point, is neither a point nor a solid (or at least dense) sphere, but a kind of huge, roughly spherical cloud whose temperature and brightness fluctuate with time.

There are a great many stars like Mira—4,566 according to the latest *General Catalogue of Variable Stars* (1969). Naturally they differ in details. Their periods range from 100 days to 2–3 years; those with longer periods have a greater magnitude range and the lowest temperatures. The spectra are generally of type M or, rarely, S, but there are also many of type C, the so-called carbon stars. In this last case, when the temperature falls to around minimum, the carbon molecules join together as carbon particles, and the stars are surrounded by thick smoke screens—red stars smoking like embers.

PULSATING WHITE STARS

Variables of the Mira type (also called long-period variables) are the most striking, but not the most representative, of a vast category of stars whose chief characteristic is that they vary continuously, with periodic pulsations like deep breaths. To this category belongs another type of variable, characterized by more frequent and regular variations. These are the so-called Cepheids, named after the leading member of the group, δ Cephei. They are giant stars, but appreciably smaller than the long-period variables; moreover, they are white and have considerably higher temperatures. They are, in short, completely different from the Mira variables, which almost certainly pulsate for different reasons.

Let us take a sufficiently typical specimen of these stars—for example, δ Cephei itself. The brightness rises rapidly to maximum, falls more slowly to minimum, then rises again to maximum, and so on, always following the same light curve. The variation is not very great (only 0.8 magnitudes) but rather rapid: each complete cycle lasts 5 days, 8 hours, 47 minutes. Continuous variations in color indicate that the temperature varies simultaneously, passing from a minimum of 5,700° K to a maximum of 7,300° K. Spectroscopic observation reveals that the spectral lines are displaced alternately toward the red and toward the violet. We have already seen that this is the so-called Doppler effect, which indicates a displacement of the luminous source along the line of sight: in the first case the source is receding from the observer, in the second approaching. It is not possible to interpret this effect as a continuous oscillation of the whole star in space. The only possible interpretation is that when we observe a radial velocity of recession, it is not the whole star that is moving away from us but only the outer layers, from which we receive the light. Hence the star is contracting. Little by little the velocity of recession diminishes, and when it finally stops, the contraction has ceased and the radius has reached its minimum value. From that moment on, we observe a velocity of approach that steadily increases, attains a maximum, and then diminishes until it stops. This means that the star is expanding with increasing rapidity, then the expansion slows down (although the star continues to expand), and finally the expansion stops: at that point the radius has reached its maximum value. After this phase the velocity again becomes positive, the star starts to contract, and the entire process is repeated (Fig. 4.8).

Measuring the variations of radial velocity may lead us to an important discovery. In fact, having observed the velocity, in kilometers per second, at which the radius of the star changed from its maximum to its minimum value, and knowing how many seconds it took, we can immediately determine how many kilometers have been traversed: in other words, we can measure in kilometers the change in the star's radius.[2]

At this point a bit of reasoning may enable us to advance another step, to the point where we can actually measure the value of the radius. Let us recall that if two equal surfaces are at the same temperature, they emit the same quantity of radiation, that is, they are equally bright. If, however, we increase the surface area of one of the sources without changing the temperature, the intensity of the light it emits will also increase proportionally. Thus by measuring the ratio of the two intensities, we can obtain the ratio of the two radiating surface areas as well. Let us return to our Cepheid. We have seen that the temperature varies periodically, as shown by the variation in color. Now let us find two different moments, over the course of the entire phenomenon, when the color of the star is the same; at these two instants the star's brightness will not necessarily be the same. Hence any difference in brightness will be due solely to the different sizes of the luminous surface at these two moments. Thus the simple difference between the magnitudes of the star at the two instants of equal temperature immediately gives us the ratio of the luminous intensities, which gives us the ratio of the surface areas and hence of the radii. Knowing the difference between the radii and their ratio, we can determine, using a simple system of equations, the two values of the radii in kilometers.

In the case of δ Cephei we find that the minimum radius is 25,800,000 kilometers and the maximum 29,350,000 kilometers. Therefore δ Cephei, though not as enormous a star as Mira Ceti, is still a giant, with a mean diameter forty times that of the sun. Taking into account the fact that the temperature is also relatively high, we conclude that it must be intrinsically very bright. And in fact it is. Indeed it turns out that all the Cepheids are very brilliant stars. Naturally not all are equally bright: some are fainter and others are up to four hundred thousand times brighter than the sun. This fact has proved to be one of the greatest pieces of good luck to come astronomers' way.

As we have already noted, even if these stars form a homogeneous

[2] Naturally, the same reasoning holds for any other phase of the phenomenon.

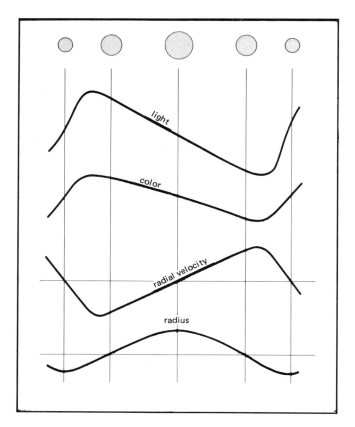

Fig. 4.8　The variations in light, color, radial velocity, and radius of a Cepheid variable.

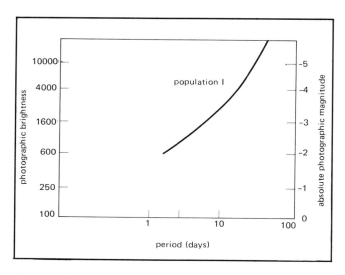

Fig. 4.9　Period-luminosity relation for classical Cepheids. Note that to a good approximation the increase in luminosity is proportional to the period.

group, they possess individual characteristics that can vary from one instance to another. One of these characteristics is the period. We know of some Cepheids (like the group of the so-called RR Lyrae-type Cepheids) that have very short periods, less than a day, and others that have periods of several days, even some tens of days. The stars in this second group, which includes δ Cephei, are known as the classical Cepheids. At the beginning of this century, it was discovered that there is a connection between the length of the period and the mean intrinsic brightness of the Cepheids, in the sense that those having a longer period are also brighter. Given this fact we can immediately deduce the distance of a Cepheid whose period and apparent magnitude are known, and these quantities are quite easy to determine.

In fact we need only proceed as follows: We construct the period brightness relation by means of Cepheids whose distances are known (that is, those for which we have the absolute magnitude as well as the period). This relation can be shown clearly by means of a graph (Fig. 4.9). We then enter on the graph the period of a Cepheid of unknown distance and note the absolute magnitude corresponding to that point. Since we know the mean apparent magnitude, we can immediately derive the distance, using a formula that we already know.

Let us apply this procedure to δ Cephei. The period is 5.37 days. Corresponding to this is a mean photographic absolute magnitude of -2.9; since the mean apparent magnitude is 4.65, we obtain (see appendix A) a distance modulus $m - M = 7.55$, that is, a distance of 1,030 light-years. Thus we now have a total picture of δ Cephei: it is a pulsating variable, more than 1,000 light-years away, sixteen hundred times brighter than the sun.

But, more important, we have discovered a very powerful tool for probing the remotest depths of space, for Cepheid variables are present in every part of the universe explored thus far, and since they are extremely bright, they make it possible for us to measure the distances of the remotest groups of stars, those for which no other method would suffice.

STELLAR EXPLOSIONS

Many centuries ago, on a July night in the year 134 B.C., humanity was struck for the first time by a strange and unexpected sight: a new star appeared in the sky. It shone extraordinarily brightly for some time, then slowly it faded and disappeared. Pliny informs us that the strange apparition led the Greek astronomer Hipparchus to compile a catalog of all the stars "so that posterity might know whether changes in the sky indeed occur." And posterity did in fact observe numerous changes like that seen in 134 B.C. Up to the present day over two hundred more stars have appeared, not all equally bright.

These stars still go by the name ancients gave them: "novae." But nowadays we know that they are not really "new" stars; that is, they are not born at the moment they appear before our eyes. By carefully studying photographs of the region taken before a nova appears, it has almost always been possible to find a faint star in its place and occasionally to determine this star's physical characteristics. It was found that a nova is almost always a star about as bright as the sun but subject to slight fluctuations. Suddenly the nova begins to increase in brightness, becoming, within a few hours, up to 150,000 times brighter than it was originally (Fig. 4.10). It is not hard to imagine what this means in practical terms. It is as if the sun were to become 150,000 times hotter and brighter within a few hours. The light would blind us; we would seek refuge from the tremendous heat in the bowels of the earth, but the respite would not last long; in the succeeding hours the oceans would evaporate, condensing in an extremely heavy atmosphere, which would in turn be dissipated in space as the heat increased.

The increase in brightness is accompanied by an increase in volume. The outer layers of the star expand at the enormous velocity of 1,000 kilometers per second. A planet as distant from the star as the earth is from the sun would be overtaken and obliterated within two days; every form of life, every trace of civilization, the planet itself—all would disappear, just as a sheet of paper, along with the words written on it and all that they signify, would be destroyed by fire.

After this cataclysm, which generally lasts two or three days, the brightness begins to decline, at first rapidly, then more slowly, until finally— sometimes after several years—it returns to what it was initially (Fig. 4.11). This second phase can be explained if we suppose that the envelope, as it is expanding, gradually dissolves, cools, and becomes steadily less bright and more transparent until, finally, the light that we observe comes only

Fig. 4.10   A nova photographed at minimum and during the explosion. This is Nova Herculis 1934 in two plates taken at the Observatory of Pine Torinese under the same conditions, as is shown by the images of the other stars. (*Courtesy of A. Fresa*)

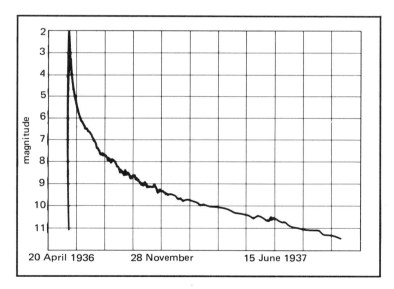

Fig. 4.11   Example of the light curve of a nova (CP Lacertae). Note the abrupt increase and the slow diminution of the light immediately after the maximum.

Fig. 4.12   The envelope expelled from Nova Persei in 1901, as it appears today, observed with the great Mount Palomar telescope. The distinct asymmetry indicates that the emission of matter during the explosion was not uniform in all directions, or else that the expansion may be partially impeded by something, for example, by interstellar matter arranged asymmetrically with respect to the center of the star. (*Mount Wilson and Palomar Observatories*)

from that lower portion of the star that was not involved in the explosion, a portion much smaller and hence fainter than the envelope expelled at the moment of maximum brightness.

The cataclysm is over. It was a kind of giant flare, in which the star dissipated, in a few months, an energy of 6×10^{45} ergs—as much as the sun emits in ten thousand years.

Meanwhile the matter expelled by the explosion continues to move away and in some cases forms around the star an envelope so extensive that it is visible from the earth (Fig. 4.12). The quantity of matter expelled is not very great, relatively speaking: according to the most recent estimates it is about one-thousandth of the solar mass. Thus if the star had the same mass as the sun, the mass expelled would be equal to about one-thousandth of its total mass. But let us not forget that even so small a fraction corresponds to three hundred times the mass of the earth.

Not all novae show such typical behavior, but the variety of types does not alter the essence of what has been said. Moreover, various stars have been observed in the nova phase more than once; indeed, according to some authorities, all novae are recurrent, but the phenomenon appears more rarely the more violent the explosion is.

This picture, which we have been able to sketch by combining the photometric and spectroscopic observations made so far, is being rapidly extended by the introduction of new observational techniques. For the moment the findings almost exclusively concern Nova Serpentis 1970, the most recent of the brighter novae. Discovered by the Japanese astronomer M. Honda on 13 February 1970, it reached its maximum visual brightness on 18 February and after a few days duly appeared to begin its slow decline. Whereas most astronomers followed the nova by traditional methods without coming across anything unusual, some followed it in the infrared. In particular G. L. Geisel, D. E. Kleinmann, and F. J. Low observed it in seven wavelengths (from 1 to 25 μ m) from 19 days to 111 days after the discovery. According to these observations the nova became, in the infrared, one of the most brilliant stars in the sky—not at the moment of the visual maximum as the optical observations suggested, but instead more than three months later. Similar results were obtained independently by A. R. Hyland and G. Neugebauer.

The infrared spectrum contained no indications of absorption, and a few calculations led astronomers to conclude that the infrared emission must come from an envelope of dust expanding at a velocity that turned out to be 800 kilometers per second, that is, of the same order of magnitude observed optically for the Nova Serpentis 1970. This envelope of dust would have been formed during the explosion of the star; it would have emitted radiation, having been strongly heated by radiation from the central star; thus, the central star must have reached a brightness greater than that shown by optical observations precisely during those days in which the latter showed a rapid decline. Moreover, the diminution of the visual brightness of novae, which according to the latest discoveries is accompanied by a rapid increase in the infrared emission, is not real but is caused by an increase in the opacity of the dust envelope surrounding the star. This in turn is due to an increase in the number or size of the dust particles. In the three months after the explosion, therefore, the visual brightness of the central star would have in fact increased by a factor of five, not diminished by a factor of one hundred, as the optical observations led us to believe. Nova Serpentis 1970 has shown, in addition to this extraordinary behavior in the infrared, an intense radio emission and a behavior in the ultraviolet analogous to that in the infrared that is equally unexpected. The ultraviolet observations have been made from space with the instruments on board the artificial satellite *OAO-2*. Clearly, we may expect to see a considerable increase in the research and discoveries in this field in the next few years.

However, neither these discoveries nor the existence of various types of novae in any way contradict the essential aspect of the phenomenon: the explosive nature of the variability. In fact, whatever the causes may be that produce these phenomena and the complexity of their development, one thing is certain: the process just described corresponds to true stellar explosions, even if only of the stars' outermost layers.

The fact that a star can explode and that the brightness of novae at minimum (that is, prior to the explosion) is roughly the same as that of the sun makes us wonder whether the sun itself might not become a nova.

We can rest assured that it will not, because the physical characteristics of stars that become novae, except for their brightness, are unlike those of the sun. They are blue stars that have masses less than that of the sun and are subject to small but noticeable light fluctuations. There is, moreover, another characteristic that makes them completely different from the sun:

they are all close double stars. This fact was discovered for the first time in 1954 by M. F. Walker, in the case of DQ Herculis. Since then, many ex-novae have been studied from that point of view and nearly all have proved to be double stars. The few that have not been shown to be double almost certainly belong to a specific category that we are already familiar with: those stars whose binary nature cannot be discovered by any type of observation.

This binary property is perhaps the origin of the explosions. In fact the considerable proximity of two stars, along with the powerful reciprocal effects on the outermost layers, might produce such unstable conditions as could only culminate in a violent phenomenon. Once this had occurred, everything would calm down again, at least for a while. If this is really the reason for the explosions, comparisons with the sun should cause us no concern.

THE SUPERNOVAE

In July 1054 Chinese and Japanese astronomers were struck by the appearance of an exceptional, extremely bright star, which they described in their chronicles. This star, which appeared in the constellation Taurus, was brighter than the planet Venus; in fact, it was so bright that for the first 23 days it was visible even in broad daylight. Its color was reddish white. After a few days of such brilliancy the brightness began to decline until, toward the middle of April 1056, it disappeared without a trace—or so those astronomers, restricted to observation with the naked eye, believed. Today, however, we have found the traces of this star and have discovered that what remains is even more marvelous than the apparition the Oriental astronomers observed. For we now know that before 1054 it was just another star; in 1054 a tremendous explosion took place, from which enormous heavenly bodies were born, whose existence, until recently, was not even conceivable.

Here is the history of the observations of this extraordinary object, begun in the Orient and, by chance, resumed in Europe seven hundred years later. When, in the middle of the eighteenth century, the astronomer C. Messier listed all the interesting celestial objects visible in the telescope, he began his catalog with a strange, small nebula in the constellation Taurus, called the Crab nebula on account of its shape. This nebula had a diffuse back-ground and a group of red filaments issuing from its central zone, branching out like coral (Fig. 4.13). In 1921 K. Lundmark noticed that this nebula was very close to the position where, according to the oriental sources, the extraordinary star of 1054 had appeared. Meanwhile, other astronomers discovered, from small changes that could be discerned in photographs taken years apart, that the nebula was expanding. They found that it must have taken about 900 years to reach its present dimensions (Fig. 4.14). There was no longer any doubt that the Crab nebula was what remained of the explosion observed in 1054. The spectrum was obtained, and from it the velocity at which the nebula was expanding was determined. By combining the value of the radial velocity, expressed in kilometers per second, with the value of the annual angular displacement, already measured on the photo-graphs, it was possible to derive the angle that the annual expansion of the nebula, measured in kilometers, subtends when viewed from the earth. From this datum it was possible to determine directly the distance of the nebula from us, which turned out to be 4,100 light-years (6,600, according to some more recent measurements). Given the distance and the apparent brightness from the ancient chronicles, it was possible to deduce the actual brightness of the star at the time that it appeared. We can imagine the astonishment of astronomers when they found that this brightness was 300 million times that of the sun. There is no known star of comparable bright-ness; even the brightness attained by novae at maximum, as we have seen, is only about 150,000 times that of the sun.

But the star that appeared in 1054 was not unique. It was found that two other stars, which appeared in 1572 and 1604, respectively, must have also attained an extremely high luminosity. These exceptional stars were called supernovae.

The name might seem to suggest that the supernovae are just unusually bright novae. In reality the two phenomena must be totally different: while in novae the explosion involves only the outermost layers, in the case of supernovae it is almost certainly the whole star that is destroyed in a tre-mendous explosion whereby it suddenly releases almost all the energy it contains. According to theoretical research carried out in the past twenty years, it is actually during these explosions that the heavy elements (such as iron and all the others up to uranium) are formed.

Fig. 4.13   The Crab nebula in the constellation Taurus. Discovered by Messier around the middle of the eighteenth century, today it has become a mine of information for astrophysics and one of the most extraordinary objects in the sidereal world. (*Mount Wilson and Palomar Observatories*)

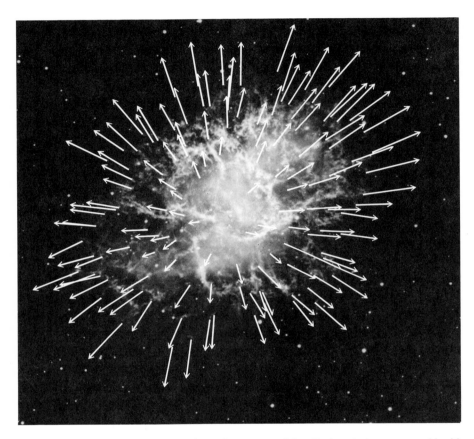

Fig. 4.14   Motions of expansion from the center of the Crab nebula measured by V. Trimble for 132 points. The arrows represent the distances that the filaments will cover in the next 270 years at their present velocities; the point is the center of the expansion deduced from the observed motions.

The discovery that the Crab nebula coincides with the supernova of 1054 was just the starting point for a series of discoveries, each more extraordinary than the one before. In 1948 it was discovered that the Crab nebula emits radio waves. A few years later the light from the diffuse background was found to be polarized. Physicists then realized that these remnants of a star that exploded about a thousand years ago were producing synchrotron radiation: an emission of optical and radio radiation caused by electrons moving at speeds near that of light in a weak magnetic field.

In 1964 the Crab nebula was also shown to be an intense source of X rays, and theoretical astrophysicists began to develop bold interpretations to explain all the observed facts. But the most unexpected discovery was yet to come.

In July 1967 radio astronomers at Cambridge University, using an instrument constructed for carrying out research on the scintillation of celestial radio sources, accidentally discovered that radio impulses arrived from a certain point in the sky at fixed intervals of 1.3 seconds. Physical considerations led to the conclusion that the transmitter must lie far beyond the confines of the solar system and that it could not be a star but must be an object not much bigger than a planet. These characteristics suggested an artificial emission, and the Cambridge astronomers jokingly referred to this "object" as LGM (for "Little Green Men"). There would indeed have had to be many little green men in various parts of space, for a while later the same radio astronomers found three more sources of this type, and today some fifty have already been identified. Because of the pulsating nature of the radio emission, these sources were called pulsars.

From the moment pulsars were discovered, two types of objects were singled out as possible sources of this strange radio emission: white dwarfs and neutron stars. The first we have already discussed. Neutron stars are bodies having a mass greater than that of the sun but very small radii, scarcely some fifteen kilometers. Their density is extremely high: from a hundred thousand to a thousand million tons per cubic centimeter. For us such values are absolutely inconceivable, just as we find it difficult to imagine the physical state of such a star.

Stars of this kind have never been observed, but as early as 1933 the astrophysicist F. Zwicky had predicted them as the product of the explosion of supernovae.

In November 1968 it was discovered that the Crab nebula contains a pulsar. The impulses were emitted at very short intervals: one every .033 seconds. An attempt was then made to find out whether they came from a body that could be observed optically. This attempt was successful.

It was not unreasonable to wonder whether, after the explosion of 1054, something might remain of the original star and whether that something would still be visible. In 1942, with this idea in mind, the astronomer W. Baade pointed to two faint stars near the center of the nebula. One of these, with an anomalous spectrum, might be the object in question. In January 1969, when this star was observed by means of special techniques, it was found to emit very intense flashes of light at the same frequency as the pulsar. In other words, though Baade's star may appear similar to other stars when it is photographed in a telescope with an exposure of a few minutes, it is not a normal star, but a source that emits light in bursts, thirty flashes a second. Between one flash and the next the star is dark and invisible, even in the world's largest telescope, but every .033 seconds it emits a flash a hundred times brighter than the sun (Fig. 4.15). In the spring of 1969 it was found, finally, that the X-ray emission also shows pulsations like those of the radio and optical emissions.

After all this—and especially having ascertained the coincidence of a radio and optical pulsar with the remnants of a supernova—the theoretical astrophysicists happily concluded that neutron stars, which they had predicted as the remains of a star after the supernova phase, really do exist.

By mid-1969 the period of rapid discoveries was over, but the theoretical research continued to develop: its purpose was to clarify the mystery of the rapid pulsation, to resolve the various problems of the relation between the central neutron star and its surroundings, and to construct the most plausible model for the structure of the neutron star itself. The interiors of these stars can only be studied theoretically because they are inaccessible to observation and because the matter they contain is in an altogether peculiar state; this state is not only incomparable to any state obtainable in our laboratories on Earth, but also to the state of matter on other heavenly bodies. Despite these obstacles it has been possible, starting from the few certain, or at least reliable, data (such as mass, dimensions, pressure, etc.), to form a fairly comprehensive picture of matter in this extraordinary state,

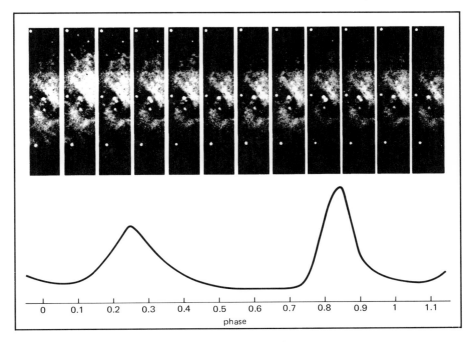

Fig. 4.15   The pulsar at the center of the Crab Nebula, photographed by means of a special process at Kitt Peak Observatory, in correspondence with the various phases. *Below:* The light curve as a function of the phase. Bear in mind that the whole period lasts only 33 thousandths of a second.

even though for reasons that will soon become apparent, we shall never be able to experiment with this matter.

A neutron fluid forms as the result of an enormous compression of matter, which occurs in certain stars having masses between 0.1 and 1.4 times that of the sun. It is the consequence of certain processes (only partially understood) in which the matter of the stellar sphere collapses rapidly toward the center. It is at such moments that the star becomes a supernova. In consequence of this, when the density exceeds 100 tons per cubic centimeter, the motion of the free electrons (and in such conditions all the outer electrons are free) approaches the speed of light. As these electrons collide with nuclei, they succeed in penetrating the latter's interior, where, neutralizing the positive charge of the protons, they form neutrons.

As the density increases, the number of neutrons with respect to the number of electrons and preexisting nuclei increases continuously until, at a density of the order of a hundred billion tons per cubic centimeter, practically all that remains is a neutron fluid, a billion billion times denser than ordinary steel and a hundred billion billion times more incompressible. This fluid can again support the enormous downward pressure exerted by the gravitational force of the collapsed star, which thus becomes a stable neutron star with a diameter of about 30 kilometers.

This star, as heavy as the sun but very small and quite strange, would appear more or less as follows, as we proceed from the outside toward the center: First of all, we would come across a gaseous envelope, less than one meter thick, composed essentially of completely ionized atoms and free electrons (Fig. 4.16). Below this extremely thin "atmosphere" is a sort of solid crust 3.5 kilometers thick. The upper part, corresponding to about half of this thickness, is made up of the nuclei of heavy elements with atomic numbers ranging between 42 and 140 (that is, from molybdenum to very heavy transuranic elements that we have never succeeded in synthesizing in our laboratories); the lower part is made up of a mixture of this upper solid fluid and the neutron fluid, which starts to predominate. The underlying zone, some 10 kilometers thick, consists solely of the neutron fluid. This "fluid" should not be thought of as something flowing, like our liquids (even those of high viscosity), but rather as matter in which the particles are not arranged in the crystalline structure characteristic of most of our solids (and,

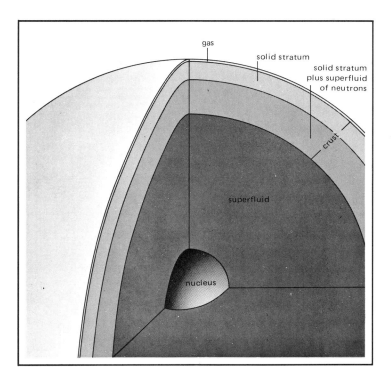

Fig. 4.16   Section of a neutron star: from the outside a thin stratum of gas (often only a few centimeters thick) is visible; there follows a crust formed by an outer solid stratum and a solid stratum mixed with a superfluid of neutrons; then comes a stratum composed of a superfluid of neutrons and protons; the nucleus is perhaps composed of a mixture of heavy elementary particles.

in a certain sense, of the "solid" matter in the upper stratum of the same neutron star). At the center, finally, we find a nucleus, 3 kilometers in diameter, in which neutrons and electrons possess extremely high energy and produce elementary particles like some of those produced by our most powerful accelerators; whereas such particles last less than a millionth of a second in terrestrial laboratories, under the conditions of temperature and pressure found on neutron stars they are stable.

What these particles are, and, more precisely, what the state of this nucleus is, not even the theoretical physicists know exactly. Moreover, the whole neutron star, even though we have just described it, eludes any possible comparison with the world known to our senses. How can we have any idea of the enormous density or of the sensations our bodies would experience on the surface of a globe where, though the globe is much smaller than the earth, we would weigh so much that an infinitesimal part of our own weight would be enough to crush us? Nearly impossible, as well, is any physical comparison of the interior of a neutron star with bodies and concepts familiar to us. In particular, the temperature of these stars, although extremely high, does not determine the properties of matter. These properties are determined by the so-called Fermi energy, which, according to quantum theory, can be linked with the density. The result is that, although the temperature must be inconceivably high, matter behaves more or less like liquid helium near absolute zero (that is, at an extremely low temperature).

The situation with regard to the outer gaseous stratum is, however, clearer. It has a very high temperature, about a million degrees, which varies with time. According to the calculations of S. Tsuruta and A. G. W. Cameron, a neutron star, which at birth is at about a million million degrees, falls to ten billion degrees in the first second, and to fifty million degrees in the next five minutes. After a year the surface temperature is ten million degrees, after 100 years seven million degrees, and after 150,000 years one million degrees. A comparison with the surface temperature of the sun (six thousand degrees) shows that even after 500,000 years the temperature is very high. The emission of energy—and hence also of light—per unit of surface area must be enormous. Nevertheless, the surface of a neutron star is so small that even a fairly recent (and thus hotter) one close to us, such as the star in the Crab nebula, is altogether invisible from the earth.

Of course, we have been discussing the light emitted from the surface. That of the periodic outbursts, however, almost certainly originates in the star's surroundings.

Recent research has led us to believe that at a distance of about a thousand kilometers from the star there is a region that emits light, radio waves, and X rays, and rotates very rapidly around the central body. This emission periodically strikes the earth at each passage, like a beam from a lighthouse. But how different this star is from the lighthouses set up along our coasts! The latter help navigators to avoid dangers at sea; the former marks the site of a cosmic shipwreck.

The idea of a shipwreck stirs our imagination: we instinctively wonder what a supernova was before it exploded. Was it perhaps the center of a planetary system, inhabited by intelligent, happy beings, which it destroyed in an instant? Or was it a star that had never reached a state of equilibrium and had never had a train of planets? Unfortunately supernovae are much more difficult to study than novae, since supernovae relatively close to us are rare: we have known only three such supernovae so far, those of 1054, 1572, and 1604, all of which appeared before the invention of the telescope. Today, with the largest telescopes, we can observe each year a certain number of supernovae in very distant agglomerations of stars that we shall reach only toward the end of this voyage. But because of their enormous distances, these stars appear to us very faint, and it is very difficult to observe them, even using the world's largest telescopes. Nevertheless, various interesting discoveries have been made.

First of all it was found that there are at least two types of supernovae. Those of type I are brighter and diminish in brightness very regularly, at first rapidly, then slowly. The average frequency of supernovae of this type is one every 450 years. The supernova of 1572 and the one that gave rise to the Crab nebula were certainly of this type.

Type II supernovae, on the other hand, are more heterogeneous. Their light fluctuation is not so typical, and their brightness at maximum is appreciably less than that of type I supernovae. According to the Russian astrophysicist I. S. Shklovskii, their frequency is appreciably higher: one every 50 years.

Other kinds of information (spectroscopic, radio, theoretical) have also come to light, but we shall not dwell on them because they are still not sufficient to enable us to discover the two most important things: what supernovae are before they explode, and why they explode.

But the idea of a cosmic shipwreck, of a cataclysm so great that it could involve all the planets of a planetary system, and perhaps those of neighboring stars as well, remains fixed in our minds. Indeed the energy liberated by the explosion is enormous: a hundred thousand times greater than the energy liberated by the explosion of a nova, itself ten million million times greater than the total energy emitted by a solar flare. And let us not forget that a solar flare is two hundred million times more powerful than a megaton atomic bomb. Moreover, we now know that X rays and cosmic rays are emitted along with the light—a much more disturbing fact. Since it is human to be primarily concerned about ourselves, we naturally wonder, "What would happen if a supernova exploded not too far from the solar system, and what is the probability of such an explosion?"

K. D. Terry and W. H. Tucker (a biologist and a physicist, respectively, both American) have tried to answer these questions by calculating the flux of cosmic rays that would reach the earth from a given distance. They then compared this flux with the doses lethal for the forms of life on the earth, and finally, taking into account the frequency at which supernovae should appear, they determined the intervals at which the event should take place in our vicinity.

The starting point is the energy emitted by supernovae in the form of cosmic rays. This value has been calculated, independently and by various means, by Russian and American scientists, and they have arrived at pretty much the same result. From this it is possible to determine what percentage of energy reaches the highest part of our atmosphere—assuming, naturally, that the star is at a specific distance. The actual flux of cosmic rays is known, and it is also known that the flux produces a dose of 0.03 roentgen a year, which is by no means enough to disturb life on Earth.

Now let us consider the type II supernovae, which seem to be more frequent. Assuming an appropriate distribution in space and supposing that the explosions occur at a rate of one every 50 years, Terry and Tucker find that, within a sphere of radius 600 light-years centered on the earth, one supernova should explode every 50 million years. In that case, assuming the supernova were 600 light-years away, the cosmic ray dose that would reach the earth within a few days would be at least 500 roentgens.

Of course, if the supernova were closer, the irradiation would be greater (assuming that the energy released in the explosion remains the same), but such an occurrence would be rarer. In practice, during the 600 million years from the Pre-Cambrian period until the present, ten supernovae capable of producing an irradiation of 500 roentgens may have exploded, four capable of producing 1,000 roentgens, and only one that could have produced a dose of 2,500 roentgens. Comparison with the present cosmic ray flux shows that these doses are very large, probably lethal. Thus the explosion of supernovae in the remotest past might explain one of the most mysterious phenomena in the history of life on our planet, that of the sudden disappearance of whole species of animals, which occurred many times in the various geological eras. In order for us to determine how acceptable such an interpretation may be, it is necessary to specify the cosmic ray doses that are lethal for the various forms of life on our planet.

Laboratory experiments have shown that, for most animals, doses ranging from 200 to 700 roentgens are lethal; moreover, a large proportion of animals are rendered sterile by exposure to radiation.

Vegetables turn out to be much more resistant. According to experiments made some years ago, the lethal dose for most plants is over 2,000 roentgens. From 500 to 800 roentgens will seriously inhibit the growth of conifers, but for deciduous plants 3,000 to 7,000 roentgens are required. Seeds and spores are even more resistant: they can withstand some tens of thousands of roentgens.

Marine life is another story. We know that the intensity of cosmic rays falls off sharply passing through a layer of water; in the sea, at a depth of 25 meters, a dose of 1,000 roentgens is reduced to only 72 roentgens. Thus plankton, fish, and marine plants living near the surface could be destroyed by a supernova explosion, whereas marine life below 50 meters could survive undistrubed, even if the surface dose were to reach 5,000 roentgens.

This selection in the distribution of living species would explain how it is that supernovae have not succeeded in destroying all life on the face of the earth, since if a few traces survived, life could have begun to develop again as soon as the destructive radiation had expended itself. Moreover, according to Terry and Tucker, the supernova hypothesis would even explain how it happened that the extinction of species occurred simultaneously over the entire earth, including the seas, and why the flora seemed to be much less affected by these crises, which decimated, or wiped out, entire species of animals. Finally, it is significant that since the Cambrian period these mass extinctions happen to have occurred with a frequency of about one every 60 million years, nearly coincident with the frequency of explosions of supernovae close enough to produce doses of 500 roentgens.

Our exploration of the world of variable stars is now at an end. We started out by learning about stars that only appear to vary in brightness, and then we discovered other stars with real, and even greater variations; finally we came upon veritable stellar catastrophes, in which enough energy is released to spread death at enormous distances.

Perhaps one day we shall see a new, very brilliant star suddenly appear, a star much brighter than the one that formed the Crab nebula, and all the forms of life on Earth will be struck dead. We will not even have time to prepare ourselves for the end, for at the very moment that the light ray enabling us to see the explosion reaches us, the deadly radiations, which travel at the same speed as light, will reach us as well. The star itself will have exploded decades or centuries earlier, for that is the time it would take the radiation, traveling at a speed of 300,000 kilometers per second, to cover the enormous distance between us and the star. During this time no one will suspect a thing; people will continue to go about their business as usual, unaware that the end of the entire species is at hand. This may be happening to us at this very moment: perhaps even as we are talking about it, projecting it into the distant future, the deadly radiation that will destroy humanity is already on its way.

5   THE BIRTHPLACE OF THE STARS

THE NEBULAE

Having witnessed the phantasmagorical spectacles of double stars and the cataclysms of novae and supernovae, we now set our sights on other goals as we set off in search of new horizons for our knowledge and new realms for our imagination. The next target is the constellation of Orion, one of the most beautiful constellations in the entire sky and one familiar since ancient times. A telescope, even a modest one, suffices to offer us an interesting view of the region we shall visit, a bright nebula known to astronomers as M 42.[1]

We have already come across one nebula: the Crab nebula, a remnant of the explosion of a supernova. The Orion nebula, however, is greater by far, brighter, and radically different in structure, and this indicates that it belongs to a category of objects completely different from the Crab nebula. The brightest part stands out in the neighborhood of the star θ Orionis, which, when observed with sufficiently powerful telescopes, is seen to divide into two groups of bluish white stars; the more numerous group contains six stars whose principal components form the corners of a trapezium. The fainter zones of the nebula are visible only through powerful telescopes or in long-exposure photographs. In a large telescope (that is, over one meter in diameter) the region presents an unforgettable sight: around the groups of θ stars, sparkling like diamonds, are spread luminous green veils of various intensities, alternating with dark zones, which help to give form and volume to the whole. Observing then, we tend at first to hold our breath, as if for fear that these tenuous veils will come undone and disintegrate before our very eyes, but the nebula remains the same even after years have gone by, changeless in its very evanescence. So we want to know more about it, that is, what it is made of and whether there are other objects like it.

The Orion nebula (Fig. 5.1) is an enormous mass of gas, very far away from us and so extensive that it takes light itself, traveling at the speed of 300,000 kilometers per second, about 30 years to traverse it from end to end. Spectroscopic analysis has shown that the gas is composed of the

---

[1] This designation is derived from the fact that it is No. 42 in the catalog compiled in the eighteenth century by Messier, containing essentially the most beautiful objects in the sky. (For further details, see appendix C.)

Fig. 5.1   The great Orion nebula in a long-exposure photograph showing the outer zones. The central part is overexposed; thus neither the structure of the nebula nor the multiple star θ Orionis can be seen. (*Mount Wilson and Palomar Observatories*)

same elements that abound on the earth and in the stars (that is, hydrogen, helium, carbon, nitrogen, oxygen, etc.). There are also heavier elements present (such as potassium, calcium, sulphur, iron); because of the particular physical conditions in the nebula, however, these are ionized and therefore show themselves in the spectrum only by means of the so-called "forbidden lines."

In addition to the Orion nebula hundreds of other nebulae are known (Figs. 5.2, 5.3), most of them farther away or less striking, but all having more or less the same composition. The gases forming them are not in themselves luminous but are rendered thus by one or more stars situated in their immediate vicinity; the stars either illuminate the gases directly or else excite them, causing fluorescence phenomena analogous to those which occur in our fluorescent tubes.

The two cases are easily distinguishable. When a nebula is illuminated by a star, the nebula's light is the same color as the illuminating star, and the spectra of nebula and star are the same. When the light causes fluorescence, emission occurs only corresponding to particular lines (for example, the hydrogen lines Hα and Hβ, the lines of singly or doubly ionized oxygen, etc.). Thus when the predominant emission is that corresponding to the Hα line of hydrogen, the nebula appears red; when light corresponding to different lines is emitted in different regions, the nebula appears to be different colors in the various areas. The blue nebulae surrounding the blue stars of the Pleiades are an example of the first case, the Orion nebula the third, whereas most of the other nebulae appear to be of the second type, that is, red.

These differences among the various types of nebulae are basically due to the diversity of the stars illuminating them. In fact, when there is a star at a high temperature capable of exciting a nebula in its vicinity, the nebula will appear as the fluorescent type; if, on the other hand, the star is at a low temperature and the light that it emits has little energy, it will only illuminate the nebula, which will appear as the reflecting type. Of course reflection also occurs when the nebula is composed of not only gas but dust. This type of composition is very common; indeed it is typical of all diffuse nebulae, such as the Orion nebula (the dust is absent, however, in other nebulae—for example, in the remnants of novae and supernovae). We naturally wonder, then, what happens when the gas and dust of a nebula

Fig. 5.2   The Rosetta nebula in the constellation Monoceros, a complex of gas and dust enclosing, in its central zone, a cluster of very bright blue stars, photographed with the large Schmidt telescope. (*Mount Wilson and Palomar Observatories*)

Fig. 5.3   Example of both a reflection and an excitation nebula: M 20 in the constellation Sagittarius, also called the Trifid nebula because of the structure of the part below. This latter portion is excited by a system of multiple stars similar to the system of θ Orionis and emits radiation particularly in the red Hα line of hydrogen. The upper part reflects the light of the bluish star at its center. (*B. J. Bok*)

are not illuminated by any star. The answer is simple: in such a case they appear as dark zones projected against the background of a starry sky or against other nebulae.

A very beautiful example is that of the so-called Horse's Head, a dark nebula also in the constellation of Orion, a little north of the region we visited. Of course the relative opacity of a dark nebula depends on its depth (that is, its extension along our line of sight) and density.

For the bright nebulae the density has been determined fairly easily from the ratio of the intensity of certain significant lines in their spectra. It has always proved to be extremely low. In the denser regions of the Orion nebula there are no more than 15,000 atoms per cubic centimeter, a value that is inconceivably low when we consider the number of molecules in one cubic centimeter of our atmosphere (27 billion billion). The density is still, however, considerably greater than that of interstellar space, in which there is hardly one atom per cubic centimeter. From the value determined for the density, and assuming an atomic weight equal to the mean of the weights of the elements forming the nebula (taking into account the proportions in which the elements are present), we find that a cubic kilometer of the nebula weighs not more than .03 milligram and that a volume equal to that of our planet would include a mass of not more than 3 tons. If, however, we also allow for the dark matter present, and above all for the enormous volume occupied by the nebula, we find that its total mass amounts to several hundred times that of the sun. This means that by concentrating that mass of gas and dust in smaller clumps we could form hundreds of stars.

The M 42 nebula represents only the most striking aspect of the wonders of Orion, and in a sense it is merely a point of departure for more fascinating discoveries. When the region is photographed in infrared light, the nebula appears much less bright, and it is possible to detect, immersed in it, a great many stars, most of them faint and concentrated around the θ star (Fig. 5.4).

When compared with one another, systematic photographs of these stars taken at intervals have shown that most of them are variable. Their brightness, however, unlike that of Cepheids or eclipsing binaries, does not vary regularly. The variations are extremely irregular: periods during which the light is practically constant alternate with periods of great unrest, in which the star may go through practically the entire range of variation within the space of a day, and other periods when the star is semistable and only small fluctuations are observed (Fig. 5.5). The range of variation is generally within one magnitude; only for a few stars does it reach or exceed two magnitudes. The reason is not yet known.

But that is not all. Careful scanning of the region has led to the discovery of other stars (in addition to the enigmatic variables), no less mysterious and even more baffling. Before we discuss these discoveries, however, we must leave Orion for a moment and return to the region of the stars near the earth in order to examine another very strange type of variable star.

FLARE STARS

The history of these variables begins on the night of 11 May 1939, when the astronomer A. van Maanen of the Mount Wilson Observatory noticed, while photographing a region in the constellation Ursa Major, that a small star showed a considerable change in brightness on two plates taken only 36 minutes apart. Although hundreds of variable stars had been identified, as far as was known not one—not even the so-called novae—showed increases of brightness capable of occurring within a few hours. Thus the phenomenon was received coolly and with incredulity, and even though van Maanen had meanwhile made another discovery of this type, the phenomenon was quickly forgotten.

Man tends to ignore anything he cannot place into the established order familiar to him. But sooner or later the facts will return to jog him awake, and reality will continue to stand between him and his plans until he decides, if not to explain it, at least to accept it.

So it happened in this case. On 7 December 1948 an astronomer at the Tucson Observatory, whose object was to determine the distance of the star L 726-8 (recently discovered by Luyten), took five consecutive exposures on the same plate and with the same exposure time. To his great surprise he discovered that the five images were not identical, as they should have been, and that the star had varied at least 2.7 magnitudes in the course of the observation. Thus the phenomenon van Maanen had announced was not only real, it took place extremely rapidly. In the case of L 726-8 it was found that the star had become twelve times brighter in a

Fig. 5.4   The central zone of M 42 in Orion, photographed at the Lick Observatory (with a shorter exposure and greater enlargement than that used in Fig. 5.1) so as to show the nebula's internal structure. *Right,* the emission in the red Hα line of hydrogen; *left,* the continuum, in which we begin to see the numerous stars crowded into the central zone, most of them varying in brightness. (*Courtesy of K. Wurm*)

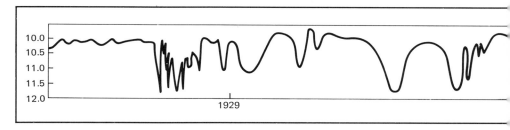

Fig. 5.5   Light curve of T Orionis according to G. B. Lacchini's observations. The abscissae are years; the ordinates are the apparent magnitudes between 10 and 12.

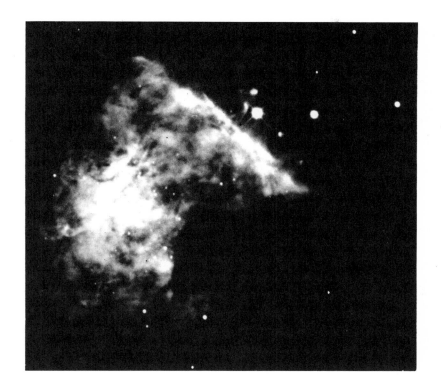

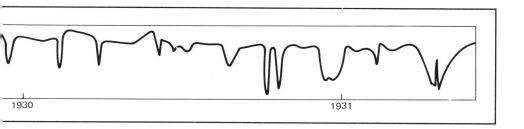

little under three minutes! This star is an old acquaintance of ours; it is, in fact, the double star that turned out to have the smallest known mass. From that point on it was no longer known by Luyten's number but by its name as a variable, UV Ceti, and indeed it has become the prototype of this new class of variables.

Some thirty variables of this type have since been discovered, scattered throughout the sky. One of these is Proxima Centauri, another star that we have already encountered.

These variables are now called flare stars (Fig. 5.6). A thorough study of them has confirmed that the increase in brightness (which can be as high as five magnitudes) is always extremely rapid, that is, of the order of 0.2 magnitudes per second. The same star can show flares of varying amplitude; a study of their frequency shows that the phenomenon repeats itself about every thirty hours and that the flares of greatest amplitude are also the rarest.

To sum up, these stars behave roughly as follows: normally they are at minimum and do not vary in brightness, at least not appreciably; at intervals of some tens of hours their brightness increases rapidly, from barely four or five times up to one hundred times the initial brightness; then it decreases slowly, and in the course of some tens of minutes the star returns to normal. Of course this schedule is not always followed strictly. The interval between two successive flares is sometimes only a few hours; the flare may be multiple, the rise or decline discontinuous (Fig. 5.7). One characteristic, however, is always present: the rapid rise to maximum.

What mechanism causes this phenomenon is still a mystery; we do not even know whether the phenomenon involves the entire star or only a particular region of its surface, for there is no star large enough or near enough to us to appear as a disk in the field of one of our telescopes, even the most powerful in the world.

One interesting characteristic, however, was discovered recently. The flares are very intense in ultraviolet light (Fig. 5.8), weaker in blue light, and barely visible in the yellow light to which our eye is most sensitive; in the red and infrared they seem not to appear at all. The study of this behavior may perhaps be a good start toward clearing up the mystery of the entire mechanism.

But we can also attempt to throw light on it by attacking the problem from

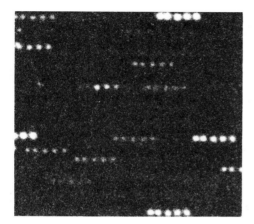

Fig. 5.6   A stellar flare observed at Asiago Observatory with the technique of multiple successive photographs. In order to find and observe variables of this type, a likely star field is photographed repeatedly with the same exposure. Thus all the stars have a certain number of equal images (in this case five) except the variables, for which one or more of the images will be brighter. Here the variable is in the center: in the first two exposures it is invisible, during the third it has risen to maximum, and then it fades in the fourth and fifth. (*L. Rosino*)

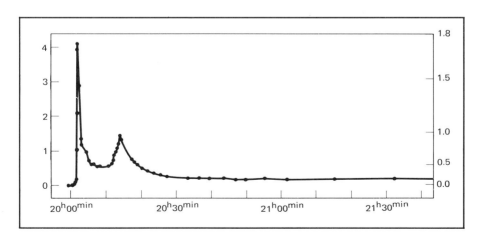

Fig. 5.7   Flares observed in the star AD Leonis at the Crimean Astrophysical Observatory on 18 May 1965. The abscissa represents the times of observation, the ordinates the magnitudes (*right*) and intensities (*left*). In less than half an hour the star shows two flares: one extremely rapid and more intense, the second slower and less bright. The evolution of the phenomenon was also followed spectroscopically.

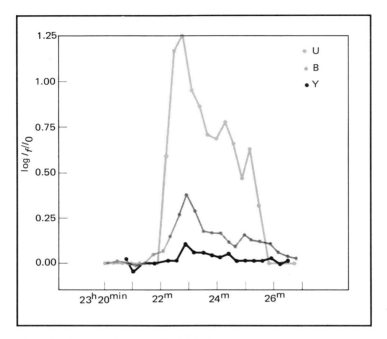

Fig. 5.8   Flare of the star YZ C Mi observed on 7 December 1975 at the Catania Observatory by S. Cristaldi and M. Rodonò, simultaneously in ultraviolet (U), blue (B), and yellow (Y). The main part of the phenomenon lasted about a quarter of an hour, but the rise from minimum to maximum took a little less than 2 min. The points in the original curves were obtained at the rate of one every 3 sec. Note the great difference in amplitude in the three colors and the complexity of the phenomenon, which showed two other minor flares: one at the start of the rise and the other immediately after the maximum.

other vantage points—for example, by trying to find out what kind of stars these flare stars are. It turns out that all of them are of low absolute magnitude (between 8 and 17) and show spectra of advanced types (K and mainly M). In other words we are dealing with red dwarfs. A property common to all their spectra is the presence of the red Hα line of hydrogen in emission.

How strange it would be to inhabit a planet moving (at the same distance as the earth from the sun) around one of these stars! Instead of the sun we should see a deep red body, a little brighter than the full moon, which every now and then would emit intense blue flares. Perhaps our eyes, accustomed to the red light, would have become particularly sensitive to red (just as here on earth they have developed maximum sensitivity to yellow light, which is that of the sun) and would not in fact see the blue flashes. All the same, the blue and ultraviolet radiation emitted by the star during the flares could substantially affect many objects and the landscape, which, from time to time and for no apparent reason, would change its appearance and color.

But now it is time to return to Orion, where a great surprise awaits us. I said that up to now only some thirty flare stars have been discovered, scattered across the sky. The region of the Orion nebula, on the other hand, is full of flare stars.

The first three were discovered in 1952. Since then their number has been increasing rapidly, along with the power of the telescopes used and the number of hours of observation. Thus the 3 flare stars known in 1952 jumped to 18 in 1958, to 42 in 1963, to 254 by the end of 1969, and finally surpassed 300 by the beginning of 1972.[2] The light curves of these variable stars are generally quite similar to those of the scattered UV Ceti type: normally at minimum, with sudden rapid rises to maximum, followed by a slow return to their original brightness. Some cases are known, however, in which the star is also an irregular variable, of the type previously described, and the flares are superimposed on the slower variation already occurring. Furthermore, of those variables associated with M 42, the rapid ones are generally of a less advanced spectral type and of higher absolute magnitude than the others. It is in any case almost certain that both groups form

[2] Nearly all have been discovered by the astronomers G. Haro and L. Rosino, in Mexico and Italy, respectively.

a single class of objects. A possible interpretation of the fact that some are isolated and others concentrated in the region of Orion—or, as we shall soon see, in other groups—will be discussed presently, when we discover what these stars really represent in the economy of the universe. First, however, we must become acquainted with a third category of stars present in the Orion nebula.

Just now we saw that the spectra of flare stars show the Hα line of hydrogen in emission. There are many stars in the region of M 42 that are not variable but show this same phenomenon. These may be variable flare stars not yet revealed as such—that is, observed, until now, always at minimum. But the possibility that they may not be variable stars has not been excluded. One thing is certain: the proportion of stars of this type in the Orion nebula is far and away greater than that of stars with the same characteristics scattered throughout the rest of the sky.

THE BIRTHPLACE OF THE STARS

To recapitulate, we have come across an immense nebula, hundreds of irregular variables or flare stars, and a great number of stars with the Hα emission line. The normal stars observed in the field of the nebula are thus a minority, and perhaps we see many of them in that region primarily as a result of perspective.

According to the Russian astronomer V. A. Ambartsumian, these three groups of stars form a physically connected whole, which he calls a "T association." Ambartsumian had likewise named the groups of stars of spectral types O and B, such as that formed by θ Orionis and by various other stars of the same spectral type scattered through the nebula, "O associations." O and T associations have been discovered in various parts of the sky. O associations are more easily visible, even at great distances, since they are composed of stars that are intrinsically brighter; T associations are visible only if they are not too distant. Furthermore, it was found that every T association is connected with an O association, whereas the converse does not hold. This effect may be due to the great distance, which prevents us from discovering the faint T association in those cases for which the bright O association is nonetheless visible. At any rate, the region of the Orion nebula contains both an O and a T association composed of the stars that we have just studied.

We must recall, at this point, that various astrophysicists have long since demonstrated that no group of stars is stable, for the individual components tend to move away from one another, dispersing in space, all the more rapidly the fewer stars make up the group. Calculations here show that the process is one-way; that is, an existing group of stars is bound to break up, but it can never happen that scattered stars join to form groups that did not exist previously. In particular, an association like that of Orion should disperse completely within about 10 million years. Hence the stars that make up the Orion association must have been born in that very region of space less than 10 million years ago. If we consider that the sun was born at least 4,000 million years ago, we immediately conclude that the Orion stars we are observing are very young; perhaps they are actually being born at this very moment. This possibility is so startling that we can scarcely believe it—we want further proof. And proof there is.

Modern evolutionary theories assert that before the internal temperature of a star is sufficiently high to permit the priming of thermonuclear reactions, the star develops energy by contracting; that is, the energy is produced by the inward collapse of matter in the outer layers under the action of the force of gravity. The greater the mass of the star, the shorter this phase.

Let us now return to the Hertzsprung-Russell diagram. The theory of stellar evolution tells us that the part we have called the main sequence represents—beyond the place in the diagram where a star is stable and thus stays for a very long time—the so-called "zero-age line," that is, the locus of the points where thermonuclear reactions begin. Stars still in the phase of gravitational contraction (that is, those stars that have not yet finished shrinking) will be found above the line, for if the temperature—and hence the spectral type—is the same, they will be brighter since their diameters are greater. The length of stay in this phase runs from 4 million years for A-type stars, which have large masses, to 700 million years for those of type M. Thus if we assume that stars of various spectral types (that is, of different masses) are born simultaneously and in equal numbers, the H-R diagram ought to show a dispersion toward the more advanced specral types since, after a certain time, the A stars would have already reached the zero-age line, the F and G stars would still lie at a certain distance above it, and the M stars would have moved only imperceptibly from the point where they started. Of course the same effect ought to be evident even if the births were not absolutely simultaneous, provided that the stars of small mass were not born so long before the B and A stars as to have reached the zero-age line before the latter were formed.

Now the H-R diagram for the stars in the Orion association shows precisely the fanlike distribution predicted by theory (Fig. 5.9). The effect was first observed by P. Parenago in 1954 and was confirmed in 1969 by M. F. Walker, by means of even more precise observations.

Thus the study of stellar evolution confirms the fact that the Orion stars are young. If the Orion association is indeed a group of stars in the very earliest stages of evolution, we should be able to find in it stars still being formed and matter about to become stars. By discovering and observing such bodies, we could then actually witness the birth of stars. A few years ago this bold research seemed to be crowned with success and, despite some contrary observations since made, is still among the most timely, fascinating areas of astronomical research.

In 1951 G. H. Herbig and G. Haro independently found three strange objects: small nebulae with starlike nuclei, located just outside the large nebula that covers the entire region. These strange objects are not the only ones of their kind; today some forty objects having the same characteristics are known in Orion and other associations (Taurus, Perseus, etc.).

In 1955 Herbig announced a sensational development. The Herbig-Haro Object No. 2, which in 1947 appeared to be composed of four nuclei, showed six nuclei in other photographs taken in 1954 with the same telescope at the Lick Observatory. Thus the two new nuclei had appeared in the course of those seven years. Examination of all available observations from 1946 to 1968 recently enabled astronomers to conclude that the two objects reached their present brightness between early 1950 and late 1952, and between late 1952 and late 1953, respectively (Fig. 5.10). The four principal nuclei (the two new ones and two old ones) turn out to be variable, and the nucleus that appeared between 1950 and 1952 has been the brightest in the group since 1966.

This discovery was discussed at length. Many astronomers were of the opinion that we had witnessed the birth of two stars, and at any rate a great

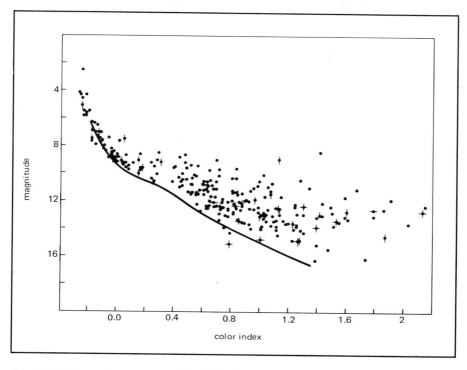

Fig. 5.9   The main sequence of the H-R diagram, constructed by M. F. Walker, for the stars of the Orion association. The zero-age line is shown in black. Most of the representative points are above it, showing that the corresponding stars have not yet completed the phase of gravitational contraction that precedes the start of the first thermonuclear reactions. The effect is more noticeable the farther one shifts toward the right, where the red stars with masses less than the mass of the sun are located; this indicates that the farther to the right the stars lie in the diagram, the slower their evolution is.

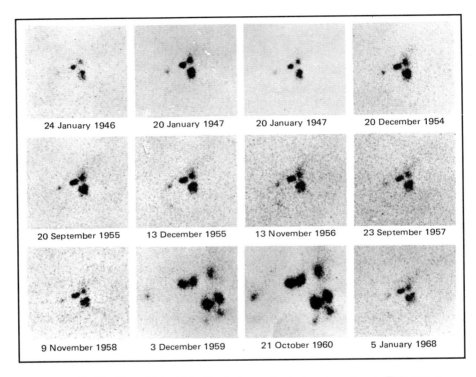

Fig. 5.10   Herbig-Haro Object No. 2, photographed at various times. This exceptional series of photographs perhaps documents the birth of two stars: the two nuclei on the left, which are visible on the 20 December 1944 photograph, but absent on 20 January 1947 and earlier. In order to show the details better, these photographs have been reproduced as negatives and greatly enlarged. The 3 December 1959 and 21 October 1960 photographs were taken with a 120-inch telescope and the rest with a 36-inch telescope, both at the Lick Observatory. (*G. H. Herbig*)

many embraced Ambartsumian's theory that the stars in question are in the very earliest stages of their evolution.

In 1960, however, Haro and R. L. Minkowski photographed these objects with the world's largest telescope and resolved them into small condensations of visual magnitudes between 18.4 and 21.5, immersed in a general, unresolvable nebulosity. These condensations did not turn out to be stars, and not one of them was shown to contain even one star. This result is in direct conflict with what we know about nebulae, particularly the fact that they become visible only if some star illuminates or excites them. In fact the Herbig-Haro objects do show a spectrum of excitation, but the photographs of Haro and Minkowski demonstrate that there is no star exciting them. What, then, makes these enigmatic objects shine? We do not know the answer. We are tempted to indulge in science fiction and to assume that the observed condensations are excited, through unknown processes, by one or more protostars contained in them but invisible to us.

Thus we come to the last chapter in this fascinating celestial novel: the search for an explanation for the facts observed. Flares, irregular fluctuations of light, nebulae that show no sources of excitement associated with them, peculiar objects—so far none of this has been explained satisfactorily. Nevertheless, Ambartsumian formulated a daring theory between 1954 and 1957. According to the brilliant Russian astronomer, both stars and nebulae originate from hyperdense prestellar matter, in physical conditions unknown to us. How that matter is transformed into the kind known to us is not explained, but let us assume that this occurs. Ambartsumian then suggests that fragments of this prestellar matter remain imprisoned in the interior of stars formed very recently. Little by little the fragments approach the surface and are transformed into ordinary matter, releasing a great burst of nonthermic energy, mostly light, which is then observed in the form of a flare. If, however, the same amount of energy is released below the photosphere, it is transmitted outward through the overlying layers, which transform it into thermic energy; in other words, the energy is not radiated directly into space but serves to heat the overlying stratum of the star, which will then show a variation in brightness solely as a result of the increase in temperature. The variation in light is smaller and slower, and the star appears as an irregular variable. Naturally, as evolution progresses, the prestellar material diminishes rapidly and the arrival of a fragment close to the

surface becomes increasingly rare. Thus, in practice, the star no longer
varies, except when it gives off those rare, sudden flashes, the flares.
According to this progression, irregular variables should be younger than
flare stars, and indeed it is found that, while the former are mostly located
above the principal sequence, the latter have already reached it and are
near the zero-age line. Even the Herbig-Haro objects and the nebulae for
which no sources of excitement are known could be explained by the presence of nuclei of this type of matter.

The theory is interesting, but it has one serious weakness: the introduction of the mysterious protomatter, which does not obey the known physical
laws and of whose existence there is no proof. Nevertheless, before we
reject Ambartsumian's theory completely, we should be prudent and at least
acknowledge that it has the advantage of reducing numerous inexplicable
phenomena to a single unknown cause. And this is in fact what Newton did
when he introduced the concept of universal gravitation.

We cannot, therefore, exclude the possibility that the existence of this
matter could become one of those fundamental postulates of science
whose validity is accepted only a posteriori, because of the ease with which,
by assuming them valid, one can explain an increasing number of other
facts.

Let us pause for a moment, content to admire all that we now know for
certain, which is no small matter. We have discovered that the birth of stars
is not an incident that occurred only in the distant past and will never occur
again, but a continually recurring event. Some stars, such as our sun, were
formed billions of years ago, but others are being born right before our
eyes. Not only that—we now know what stars look like when they are being
born and in what part of space they originate. We have discovered the
forges where they are cast, and someday, by collating and interpreting all
we see, we shall even perceive how and at what moment they are born. In
order to reach this point, we must continue to observe carefully those regions accessible to our investigation despite their distance. Some fifty star
nurseries have been found so far. These are the closest to us, and yet their
distance is so great that only some can be observed to the best advantage.
One of the closest stellar nurseries is in fact that of Orion. According to
measurements computed by K. A. Strand, its distance from us is 1,950
light-years. This figure shows that the Orion association, though one of the

closest, is at an enormous distance from us; indeed it represents the remotest part of space we have reached thus far in our journey.

We shall therefore pause a moment longer in this region, in order to look around, to observe the sun and the earth, and to find our bearings for the new regions to which we shall advance immediately afterward.

TWO THOUSAND YEARS AGO

It is winter of 1970. The last rays of a Roman sunset spread a pink patina over the three surviving columns of the temple of Mars in the forum of Augustus. The three Corinthian capitals, having once witnessed and participated in the grandeur of another age, seem to contemplate the ruin that lies beneath them. The temple was built by Augustus after the battle of Philippi, and at the beginning of the Christian Era, nearly two thousand years ago, it had just been completed and appeared in all its glory. The three columns had a great many companions and together they enclosed a cella in which a precious relic was kept: Caesar's sword. On either side of the temple, farther down, gleamed two basilicas, dazzling in their pure white marble and adorned with splendid polychrome marbles. The inner columns, of yellow marble, marked the sides of two porticos containing niches with statues of the most illustrious leaders, from Aeneas to Augustus. The entire complex was admired by a noisy crowd that would frequent the basilicas or flock to the forum from the nearby Suburra district. All that remain today are ruins or traces. Meanwhile the sunset has vanished, and the shadows of the centuries blend with the darkness of night.

A few hours later an astronomer focuses a telescope on the constellation of Orion to observe M 42. The center of the field is dominated by the multiple star θ Orionis. Around it spreads the nebula, decreasing in brightness toward its extremities. Slowly the instrument follows the object, and the astronomer prepares to receive, through the images, a message that left 1,950 years earlier, when the forum of Augustus, just built, and dominated by the temple of Mars, was seen in its entirety. At that time the ray of light began its journey. The Roman Empire had only just risen, and Christ was about to start preaching. Centuries passed, Christianity triumphed, the Roman Empire fell, the barbarians descended and destroyed millenarian civilizations, and the light that today shows us those stars continued its journey, devouring space at the speed of 300,000 kilometers per second. When Leo III crowned Charlemagne, it had not yet reached the midpoint of its journey. In Dante's time that point had been passed, but the goal was still far off. On Earth Columbus discovered America; Copernicus, Galileo, and Newton founded modern science; the French Revolution broke out; Napoleon's star shone and waned. And still the journey of the light that had left the Orion stars continued. Later the world was devastated by two wars, and at the end of the second the first atom bombs exploded. Only some years later did the ray of light reach the region of the nearby stars that we have already visited, and one night in 1970 it reached the eye of the astronomer to whom it finally showed events contemporary with Augustus and Christ.

Thus today we see in Orion not what is happening now, but what took place nearly two thousand years ago. Some of these stars may have disappeared; other, new ones may have been born. Whatever is happening there today, no one on Earth will be able to find out for two thousand years.

Nevertheless we do not suppose that, generally speaking, very radical changes could take place in so short a time, that is, short on a cosmic scale. There will undoubtedly be local changes, like those observed in Herbig-Haro Object No. 2, but the association as a whole cannot evolve to such an extent as to deviate that much from its present appearance, or at least from a foreseeable configuration.

So we must resign ourselves to our fate; we must adjust to the fact that we cannot view simultaneously all that we see. When we look at the sky, even though it may not seem that all the stars are equally distant from us, we never have the sensation that what we are seeing is not happening simultaneously. The sensation we do have is the result of our habits. If, looking out the window, we see a group of children playing in the garden, someone shaking a rug on a balcony across the street, and an airplane flying in the distance, almost on the horizon, then we think of the three events as occurring simultaneously, that is, all three are occurring in the moment that we see them, and in fact this is so. But what holds true in rough approximation for short distances is no longer valid for heavenly bodies. Because of the delay due to the time the light takes to cover the distance, we see the moon as it was a second earlier, the sun eight minutes, the planets a couple of hours, the closer stars a few years, and those less close as they appeared hundreds or thousands of years ago. We are, in short, in the position of someone who in the same day receives four letters, but one from the same city, another from a neighboring city, the third from another continent by air mail, and the fourth from that continent by surface mail: although the letters arrive at the same moment, they bring news relating to four different dates at considerable intervals.

But let us return to M 42 and allow the same imagination that transported us in an instant to a planet of Alpha Centauri to carry us in a moment to Orion—not within the association (where the bright clouds, and especially the dark ones, would hide most of the more distant heavenly bodies from us) but somewhat outside: for example, to a planet of a star some 50 light-years from the nebula. Here the sky is truly unrecognizable. True, we can still see many of the stars visible from the earth, but we would be able to recognize them only after prolonged observations: the positions are changed, the relative brightness is greatly altered, and, moreover, the large number of stars that have disappeared and the equally large number of new stars are such that we can say that we are truly under a different sky. The one familiar zone on which we can rest our gaze is the belt of the Milky Way, but even this only in its background structure; for now the stars projected against it, and even some dark nebulae such as the Cygnus nebula, are no longer those we saw in the Milky Way from the earth.

This new sky presents a magnificent sight. The Orion nebula, a marvelous interlacing of pale greenish lights, shading off in various colors to the red of the outer regions, dominates the sky to an extent sixty times greater than that of the full moon. At its center, gathered in a very narrow zone (hardly half the width of the moon), the six principal stars of the group θ Orionis sparkle brilliantly, with a brightness at least equal to that of the brightest star in our sky. Here and there, throughout its full extent, a great number of fainter stars seem to give it a spark of life. Nevertheless, though so close, we do not yet see the cluster of faint infrared stars shown on the photographs taken with our telescopes. These stars are so faint that, even after we have moved 1,900 light-years closer to them, most of them remain below magnitude 8. As compensation, we see in the region other, very bright stars and other nebulae we have not yet discussed: the nebula NGC 1977, containing ten or so very bright stars; the zone north of the dark Horsehead nebula, with the nebulae NGC 2023-2024, and the star ζ

Orionis. In short, the entire zone north of M 42, rich with bright and dark nebulae, with bright stars and strange objects, appears very striking from our observatory, in such a new perspective and with foreshortenings so different from the kind we are accustomed to that only with difficulty can we recognize the more familiar regions and stars. South of Orion, on the other hand, disappointment awaits us. Sirius, the star that, along with Orion, dominated our sky, has disappeared. We left it behind us soon after we began our journey from the earth to M 42, and we now look for it in the opposite direction, among the fainter stars not visible to the naked eye. In that same region of the sky we can also find the sun. Now at magnitude 13.5 (thousands of times fainter than the faintest stars visible to the naked eye), it is visible only through a good telescope, and even then it is lost among all the other faint stars present in the field.

Endeavoring to locate the planets that accompany it would be an arduous undertaking. The earth, in particular, being one of the smallest planets and one of those closest to the sun, would be among the first to be lost in its rays. But the same imagination that enabled us to transport ourselves in an instant to a planet not far from Orion could also put at our disposal a telescope capable of showing the earth in all its details. With this telescope we should see our planet with its blue seas and azure clouds; the cloudy white icecaps of the polar regions; Africa mostly reddish because of its deserts; the Mediterranean Sea; and the Italian Peninsula seeming to hurl itself into the sea. The impression is quite similar to the many fascinating photographs that the astronauts took on their journeys to the moon.

Let us increase the magnification and observe a particular region—for example, central Italy. We shall not have any difficulty recognizing the signs of human life: variously colored zones corresponding to different types of agriculture; streets or canals, like very fine threads; and various towns and cities. We are particularly struck by one of these cities located not far from the Tyrrhenian Sea: rather large and transected by a river, it is the city of Rome. Let us increase the magnification even more, so that we may see more detail. We are looking at the center of the city and cannot believe our eyes. We cannot find the Victor Emmanuel monument (and one could hardly call it unobtrusive!) or the Palazzo Venezia. We look around carefully and cannot even find the Colosseum. We cannot understand it: the fact that we moved instantaneously in space cannot have caused everything we saw only a few hours ago to disappear.

In reality nothing has disappeared: what we are looking for has not yet appeared. We have seen that the astronomer who photographed Orion in 1970 was in fact noting events that took place 1,950 years earlier; so now, conversely, as we look at the earth from a planet 50 light-years away from Orion, we are witnessing what happened 1,900 years ago. And if we look closely, we shall see in lieu of the Colosseum a small lake, flanked by the enormous Golden House of Nero, stretching from the Oppian Hill to the Palatine, and in lieu of the Victor Emmanuel monument the Temple of Juno, opposite the Capitoline Temple of Jupiter, the most sacred temple on the most sacred spot in Rome: the Capitol. Thus the history of some two thousand years ago is taking place right before our very eyes.

But that is not all: from the observatory in which we now imagine ourselves to be, we should be able to observe the entire history of the past two millennia unfolding. Within nine years we shall witness the eruption of Vesuvius and the destruction of Herculaneum and Pompeii; subsequently we shall see Rome in all its glory, and thus, century by century, we shall be able to witness all the events of history up to our parents' infancy and our own. Of course we shall not live long enough to reach that moment, but we need only travel toward the image, approaching the earth at an extraordinary speed, and the entire history will unfold at an accelerated pace, depending solely on the speed of our motion.

All this is pure fantasy, for physics tells us that it is impossible to exceed the speed of light, and we cannot materially transport ourselves to a distant place to await a light message that left before we did. In other words, we cannot instantaneously reach that planet near the Orion nebula, as we have just done in imagination. Thus we cannot see Titus or Trajan or the splendor of imperial Rome. That is a privilege reserved for the highly evolved beings now inhabiting that distant planet—provided, of course, that they exist.

GROUPS OF STARS

When we were exploring Orion, infrared photography disclosed to us an agglomeration of faint stars in the central region of the nebula. Groups of stars are not exceptional: it is possible to discover very many even by means of normal visual or photographic observations, and indeed some of the brightest are visible to the naked eye. The most striking group is undoubtedly that of the Pleiades, well known to anyone familiar with the night sky and also mentioned by the earliest writers. The Bible refers to them in the Book of Job, and Homer mentions them in the Odyssey when he says that Ulysses, sailing from the island of the nymph Calypso, steered his raft "seated at the tiller, with great skill, looking . . . toward the Pleiades . . . keeping the Bear always on his left."

The sight of this group of stars in the telescope is exciting. A modest telescope with low magnification or even opera glasses are sufficient to show the six principal stars—those easily visible even to the naked eye—as so many diamonds sparkling against the dark background of the night sky. Around them, tens of other fainter stars increase the brightness of the region, enhancing the beauty of the sight like the smaller stones of a diadem. The fact that these stars appear as a group is due not to perspective, nor to chance; they constitute a physically associated whole, which astronomers call an open, or galactic, cluster.[1]

Their physical connection is demonstrated not only by certain similarities of composition and structure, as we saw in the associations, but also by an extremely interesting dynamic proof that is worth noting.

The fame of the Pleiades cluster was not limited to antiquity; it has been one of the celestial objects most studied in recent times and perhaps the object most photographed (Fig. 6.1). The earliest photographs go back more than a century. Thus if any of the stars that compose it had moved in space to a noticeable extent, we should have been able to note—by comparing a photograph taken in the last century with a recent one—that they do not occupy the same positions on both photographs. Obviously, such an undertaking is rather difficult. In fact, although stars certainly move at high speeds, to our eyes they seem to remain motionless in the same positions

[1] The qualifications "open" and "galactic" will be explained later.

in the vault of the sky, which we see moving daily as a rigid whole, solely as a result of the daily rotation of the earth on its axis. For this reason the ancients called them fixed stars and pictured them, more or less (in the words of Don Ferrante in Manzoni's "*I promessi sposi*") "so many pinheads stuck in a pincushion."

The fact that obvious displacements are not observed is due solely to the enormous distance, which retards the apparent motion, just as an airplane at a high altitude appears to fly more slowly when viewed from the ground. Nevertheless, for stars moving more rapidly, and above all for those closer to us, the displacement in the sky can be discovered and measured by comparing photographs taken a long time apart. The effect is that much more evident the longer the interval between the plates. Indeed this is one of the reasons why photographs of celestial objects increase in value and importance as time passes.

Thus it was found that the star which appears to have the greatest velocity is Barnard's star, which shifts 10".30 every year; this means that in 190 years it moves through the same angle subtended by the diameter of the moon. Similar displacements, which astronomers call "proper motions," have been seen in a great number of other stars; in fact some of these had already been discovered before the invention of photography, simply by comparing the positions given in catalogs compiled at various epochs. It was then possible, by studying the trajectories, to reach some interesting conclusions. We need only recall that, as we have already seen, it was possible to deduce, simply from the wavy shape of some of these trajectories, the presence of unseen stellar companions; indeed one such companion was found near Barnard's star.

To return to the Pleiades, the possibility of examining very early plates enabled astronomers to measure the proper motions of the stars in the cluster. It was thus discovered that they all move in the same direction and at the same speed (Fig. 6.2). This is a very significant proof of their physical connection. It was also found, by measuring the proper motions of all the stars in the field, that some stars mixed in with the group move differently than the others and hence must be considered as being close only in perspective, whereas others distant from the cluster's center and at first confused with the background stars appear to move together with the group and thus form part of the system. The cluster thus appears more scattered, more open, than it seemed at first. The total number of stars contained in it turns out to be nearly three hundred, most of them concentrated in an area sixteen times that of the moon. Its diameter is about a dozen light-years. The brightest stars, known since antiquity by the names of the seven daughters of Atlas (Alcyone, Electra, Maia, Merope, Taygete, Celaeno, Asterope), are enveloped in nebulae of the reflecting type, which in color photographs appear to be blue, as are the stars that illuminate them.

The Pleiades turn out to be not that far from us—barely 250 light-years. Other clusters, such as the Hyades and the cluster of Prasepe, are even closer. Nearly all the others, however, are more distant.

Today 867 clusters are known, but the total number is thought to be much greater. They vary a great deal (Figs. 6.3, 6.4); some are quite meager, made up of a few tens of stars; others, like the very rich double cluster of Perseus, are very abundant, containing a thousand or so. The most abundant are also the brightest and, in general, the most extensive. The diameters vary greatly; the majority, like the Pleiades, measure around a dozen light-years across; some, however, are as little as three light-years, whereas others are as much as sixty. The overall color varies from one cluster to the next and is linked with the total absolute magnitude. The brightest cluster (MGC 457) has a total absolute magnitude of −9 and is bluish; the faintest clusters, with an absolute magnitude of about 0, are reddish.

All these data, relating to the clusters' volume, color, brightness, and degree of concentration, are very important for a detailed study of these stars, and we shall soon see a part of what they reveal. Meanwhile they demonstrate the great variety of these objects. All the same, what strikes us even more than the clusters' characteristics is their very existence. They prove, in fact, that in our voyage in the sidereal world we shall come across not only stars scattered irregularly (as occurred in the vicinity of our sun and generally occurs elsewhere) but also groups of varying sizes in which the stars are crowded together. In these areas the stars are closer to one another than are those in our neighborhood, and it follows that space is more crowded. The open clusters are only the first stellar agglomerations we shall encounter. Before long we shall be meeting other, much grander ones, and little by little we shall perceive the stars and all the other heavenly

Fig. 6.1 *Above,* the Pleiades, photographed by amateur astronomers in Bologna with a small instrument and short exposure, appear as a cluster, as was already shown by observations with the naked eye. *Below,* the stars of the Pleiades, photographed with a long exposure and more highly enlarged, appear surrounded by blue nebulae of the reflection type and of the same color as the stars whose light they reflect; these nebulae were discovered photographically in 1885, without ever having been visually observed. It was thus shown that the photographic plate can reveal new objects by virtue of its greater sensitivity to blue light rather than yellow, which the eye is more sensitive to. (*L. Baldinelli: Mount Wilson and Palomar Observatories*)

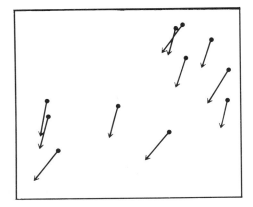

Fig. 6.2   Proper motions of the Pleiades: these stars all move in nearly the same direction, as a comparison of Bradley's 1755 observations with those of Bessel in 1840 had already revealed. The displacements of only a few of the brighter stars are shown here.

Fig. 6.3   The clusters h and χ in the constellation of Perseus, photographed at the Asiago Observatory. These clusters are made up of a great many very young, very bright stars, which appear considerably concentrated given the great distance of the two groups from the solar system, estimated at about 7,800 light-years. (*Courtesy of L. Rosino*)

Fig. 6.4   The cluster M 67. Since it is closer (a little over 3,000 light-years away) than the clusters of Perseus, it appears more scattered and is likewise made up of very bright stars, even though its brightest stars are actually between 1,000 and 10,000 times fainter than the O and B stars of the double cluster of Perseus. (*Mount Wilson and Palomar Observatories*)

bodies that we shall encounter combining in ever more abundant and immense agglomerations, forming, as it were, the successive steps of a staircase so vast we cannot discern the top. Though we have a long way to go, and even though we do not know when—or even whether—we shall be able to reach the last step, let us at least climb the first.

OTHER GROUPS: OTHER STARS

Now that we have an idea of those abundant agglomerations of stars that form the open clusters, let us continue our journey toward another wonder of the sidereal world. We can picture ourselves approaching it blindly, by imagining, for example, that the spacecraft transporting us is being piloted from a terrestrial base. Even in space it is sometimes pleasant to allow ourselves to be driven so that upon arrival we may enjoy the thrill of the unexpected.[2] Our observation base will still be a planet that physically resembles Earth in motion around a star like the sun, and has the equivalent of terrestrial days and years. When we receive the signal that the landing has taken place, we leave the spacecraft and let our eyes adjust as we slowly scan our surroundings from ground to sky. The view, though unusual, does not impress us greatly. We find ourselves beneath a sky illuminated by an intense light, like that of dawn a few minutes before the sun comes up. But this light is less yellow than that of earthly dawns; it is decidedly milky white, like the light of a lamp with opaline glass. Immersed in that light we can descry many stars, but they are dim, the way the brightest bodies in the terrestrial sky (Venus, Jupiter, Mars) appear in the predominant light of dawn. The light is diffused equally over the entire sky, and nowhere along the horizon do we discern that area of more intense light which precedes a very bright body about to rise. In fact, as time passes, the light neither increases nor diminishes; it remains the same, just as on Earth the darkness of the night sky remains unchanged as the hours

[2] Obviously this is a fantastic hypothesis, unrealizable if for no other reason than the fact that the signals linking our spacecraft with the earth can travel, at most, at the speed of light, and our present goal is at a distance of several thousands of light-years from the solar system.

go by. But in the same way night comes to an end, here also, in the strange twilight of this distant planet, a certain change occurs. Preceded by a rapid imperceptible increase in brightness along the horizon, the sun rises. The change is not very great. The light takes on a warmer tone, and the last traces of the stars disappear. On the whole, however, the illumination has not changed appreciably; it has multiplied barely a thousand times—comparable to the change on Earth between early morning and midday. The hours pass; the sun traverses the sky in its daily arc and sets unostentatiously. Before long the sky appears bright and milky again, studded with faint gleams, as we have already seen. Let us wait a while longer. Time passes, and when the sun finally reappears, we no longer have any doubt: this strange twilight sky is indeed the night sky. We have landed on a planet where there is no night—or, rather, where the night is almost as bright as day. What we have witnessed, however, is not "white nights", as occur in some regions on Earth at certain times, when the sky remains lit up by the sun, which remains just below the horizon or does not quite set. No, here the sun is no longer present; the light that illuminates the sky comes not from the region where the sun has set but from the entire sky, and it remains constant until the sun rises again. Hence, this light can only derive from the stars in the sky—but what kind of stars are these when there are so few of them (fewer than those in the terrestrial sky) and they appear so feeble? To obtain an explanation we must ask the terrestrial base where we have landed. The reply is brief: on a planet near the center of M 13.

These few words suffice to explain everything. On Earth, if we look at the sky on a moonless night, we see, between the constellations of Lyra and Corona Borealis, the extensive constellation of Hercules. On one side of the trapezium that forms the body of the mythical hero, we notice, even with the naked eye, a kind of nebulous, very faint star. Binoculars or a spyglass show it as a bright round speck, standing out clearly against the dark background of the sky. But a telescope, even only a medium-sized one, reveals an indescribable sight: against a vast milky background, decreasing in brightness from the center to the periphery, we see projected a swarm of stars, myriad sparks of varied intensity and color (Fig. 6.5). We are astonished. We have already seen clusters of stars, but this is far more crowded than even the most abundant open cluster. Furthermore, as long-

Fig. 6.5   The globular cluster M 13 (NGC 6205) in Hercules: barely visible to the naked eye as a faint luminous speck, it is actually an extremely abundant agglomeration of stars 25,000 light-years distant from the earth. (*Mount Wilson and Palomar Observatories*)

exposure photographs show even more clearly, the milky background is itself made up of stars so faint and so crowded together as to be indistinguishable, and all the stars are distributed according to spherical symmetry and become more numerous the closer we get to the center. This cluster (which appears as No. 13 in Messier's catalog) was the first of its kind to be discovered: Halley found it in 1714, but he never suspected that this speck is actually an agglomeration of stars. Even Messier had no idea. Indeed fifty years later he described it briefly with these words: "A nebula which, I am sure, does not contain stars. Round and brilliant; center brighter than the edge." The first astronomer to become aware of its complexity was W. Herschel, who was, moreover, the first to use large telescopes. He not only resolved it but tried to count the number of stars, which he estimated to be about fourteen thousand. Ever since then this cluster has been increasingly observed, photographed, and studied. According to the most recent research, it has an enormous diameter of 160 light-years and is composed of five hundred thousand stars. Its distance from the earth is 25,000 light-years—the greatest distance we have reached thus far in our journey.

But M 13 is not the only cluster of this type. Several others have been recognized for some time, and today the total number of those known—nearly all of them more distant than M 13—amounts to 121. The stars in these other clusters are also generally distributed according to spherical symmetry and are more crowded in the central regions. Because of this spherical aspect, which makes them quite similar to one another, all these clusters are called globular clusters. The distinction between open and globular clusters is based not only on this morphological criterion (which may even be lacking in certain borderline cases) but on other criteria as well, principally the dimensions, number, and physical characteristics of the stars.

The characteristics of M 13, which we have already examined, are fairly representative. Nevertheless we can get a complete picture of this type of object by considering the extreme cases as well. We find that the smallest globular clusters have diameters of 5 light-years, like those of the majority of the open clusters, whereas the largest reach diameters of 390 light-years. The cluster ω Centauri, located in the Southern Hemisphere, has a diameter of fully 620 light-years, but it is an exceptional case.

Estimating the number of stars is rather difficult. In fact simple counts will not solve the problem, because in the more abundant central zone the stars blend into one another, and as the power of the telescope is increased, their number increases to the point where they form a continuous, unresolvable background. Moreover counts in the peripheral regions have shown that the fainter the stars are, the more their number increases. Clearly this number cannot increase to infinity, and at a certain point it must start to diminish, but nothing leads us to believe that our observations have as yet enabled us to reach that point. On the contrary, at the distance of M 13, for example, with the largest telescopes in the world, we can detect stars with an absolute magnitude of 4.3, and we are well aware, from our study of the neighborhood of the sun, that there exist red dwarf stars down to an absolute magnitude of 16.7 (that is, ninety thousand times fainter). We also know—still from our study of the sun's neighborhood—that it is these very stars that are the most numerous. If this is also the case in the globular clusters, the number of stars they contain must be enormously high. On the basis of these and other considerations modern specialists have succeeded in calculating that the most meager globular clusters contain about a hundred thousand stars, and the most abundant around ten million. Recalling that the known clusters number over a hundred, we find that the total number of stars contained in them must exceed a hundred million. This staggering figure must still be considered less than the real number, since there are undoubtedly many clusters that have not yet been discovered.

But the globular clusters are astonishing not primarily because of their size, the number of stars they contain, or their high concentration. The most sensational discovery concerns the very nature of the stars they comprise, which do not resemble those we have encountered so far in the neighborhood of the sun, in Orion, or in the open clusters. Just as a Chinese and an Ethiopian differ in the color of their skin, their features, stature, and so on, though both belong to the human race, the stars in the globular clusters have basic similarities to other stars but differ in certain details. In short they are another race.

First of all, their spectroscopic behavior is strange. The spectra of globular clusters show the dominant characteristics of the brightest stars they contain. As was to be expected, they can be assigned to the spectral classes that we already know, but generally they do not contain the metallic lines, or else these appear very weak. Hence the stars in globular clusters are low in metals.[3]

But that is not all. When we visited the neighborhood of the sun, we constructed a diagram using the brightness of the stars and their spectra (or color). This was the famous Hertzsprung-Russell diagram, which shows how the brightness of the stars varies with temperature and which enabled us to discover the dwarfs and giants. Let us try now to construct the same diagram for a globular cluster. This undertaking will require patience, but it is not impossible. The absolute magnitudes can be obtained directly from the apparent ones if the distance of the cluster is known. Obtaining the individual spectra of the stars, on the other hand, would be an arduous or impossible task because of their faintness. We can compensate for this, however, by determining the color. We need only photograph a cluster twice: once with a blue-sensitive plate, and once with a red-sensitive. The blue stars will be more prominent on the first plate, and the red on the second. From the difference in intensity on the two plates we can derive a number that will express the color in all its shades.

We can then construct H-R diagrams for one of the groups of stars that we already know and for the stars of a globular cluster. They appear markedly different (Fig. 6.6). In the globular cluster the group of blue giant stars (upper left) is lacking; in exchange there are a great number of red supergiants (upper right). At any rate, the whole shape is completely altered. We can only conclude that we are dealing with stars of two different races. This was undoubtedly the impression that the astronomers had from the moment the difference was discovered—in fact they called the stars already known Population I and the stars in the globular clusters (which we shall also find elsewhere) Population II.

The differences between the two populations are now attributed to their different ages. The picture of the universe that we are tracing is not static, it reflects situations at various times, but the objects we are observing and describing do not always remain the same. Everything is changing and

[3] A certain number of clusters constitute an exception. This discovery does not alter the conclusions we shall reach; indeed it supports them. Nevertheless we shall disregard this group of clusters in order to avoid an area that lies outside the one we set out to explore.

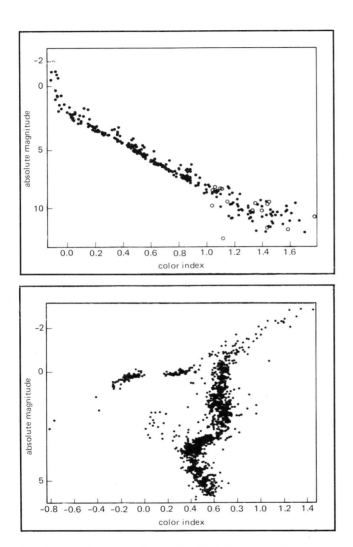

Fig. 6.6   Difference between the H-R diagrams of an open and a globular cluster. *Above,* that of the Pleiades (open); *below,* that of the globular cluster M3. The abscissae are color indices, the ordinates the absolute magnitudes. In the lower diagram the magnitude scale is reduced to half that of the upper diagram, and the color index scale is shifted to the left so that it coincides with that of the upper diagram.

evolving, and bodies formed earlier, or subjected to a more rapid evolution, may have already reached a phase beyond the one reached by other bodies. Hence the modern interpretation is that the stars of Population II are older than those of Population I. In other words the H-R diagram of the globular clusters is now what the diagram of the open clusters and the stars near us will someday become. The absence of metals tells us, then, that the globular clusters were formed in an extremely remote epoch (before the stars we have encountered so far were born), when metals did not yet exist. Hence the globular clusters are aggregates of stars formed from different material; their stars are much older than the sun and the stars of Population I, and, unlike these, they have been in existence not for a few billion years but for about 1,000 billion years.

Unfortunately we cannot dwell on this fascinating subject any longer, since our voyage would change from one in space to one in time, and a voyage in time, in search of the beginning and end of the heavenly bodies, would be not only more difficult but, above all, much longer.

A LOOK AT THE COSMOS

Let us return now to M 13, and, having found out so much about globular clusters, try to come up with a better explanation for the strange sky we observed upon our arrival. I said that the planet on which our spacecraft landed is located within the cluster, close to the central region. In that region the stars are very crowded. It has been calculated that there are at least fifty in each cube having sides of 3.3 light-years. To compare that density with a distribution familiar to us, we need only recall that in the neighborhood of the sun, in a volume a thousand times larger, we find only one star. Thus the distance between one star and the next, in the central region of M 13, is hardly 10 astronomical units—about the same as the distance between the sun and the outlying planets of the solar system. For this reason the brightness of the brightest stars among those closest to the planet falls somewhere between that of the full moon and that of the quarter moon. (To form an idea of how bright a sky containing so many brilliant sources must be, we need only recall how many stars disappear from our sky on a moonlit night.) Although the distances of these stars from the planet are small, they are still large enough for the stars to appear pointlike, like those in our sky;

and the diffused light, though considerable, is not sufficient to extinguish the brightest light sources, which grow dim one by one but never disappear completely.

Thus we have managed to explain the mysterious brightness of that "night" sky. But we were able to do so only because we had seen the cluster from outside. Such an undertaking would have been much more difficult if we had been born and raised on the planet. Its inhabitants—if there are any—would have a hard time discovering the existence of the stars, having to rely on the observation of their sun and the brighter stars that do not disappear completely. If they succeeded in constructing spaceships and traveling beyond their atmosphere, they would be struck by a stupendous sight: hundreds of thousands of stars, of various colors, shining like beacons, and far in the distance a kind of luminous dust. They would then experience fully the splendor of the universe, but they might also conclude erroneously that their cluster is the entire universe—an admissible and forgivable error. In fact, imprisoned in a world of light, these beings would be unable to explore more distant space, and perhaps they would not have even attempted the excursion outside their atmosphere that we just imagined. Or perhaps, encouraged by the relative proximity of the stars to one another and by the proximity of possible planetary systems, they have been more motivated than us to explore the nearby heavenly bodies, have reached them, and have succeeded in founding one of those marvelous interstellar civilizations we read about in science fiction novels. But even this would not have sufficed to afford them a more extensive view of the universe or to reveal other objects. To liberate themselves from this kind of geocentrism, they would have to undertake a very lengthy voyage in space. They would have to cover a distance of about 70 light-years to reach a planet belonging to one of the peripheral stars of the cluster.

Let us imagine that we are following them, for we shall also find the view of space from that region of much greater interest than the preceding view. Let us suppose that we land on a planet of a star outside the cluster when the planet is at a certain point in its orbit. During the day the landscape is illuminated by that sun, but during the night we shall see the enormous cluster in all its splendor, 90° in diameter, standing out against the background of the sky. Over an area half that of our nocturnal hemisphere we shall see more than ten thousand stars shining, some colored and extremely bright, others fainter, fading toward the center in a swarm of barely visible lights. Over the entire background a veil of light—like the Milky Way, only brighter—indicates the presence of the faintest stars, which the naked eye cannot discern individually.

Months pass. As the planet shifts in its orbit, the cluster will no longer be visible, since it is in conjunction with the sun and therefore above the horizon in daylight. But at night we shall be struck by an even grander sight. The sky is almost completely dominated by an enormous, slightly compressed wheel, 120° in diameter, bright as the Milky Way, strewn with an infinite number of small stars, bright nodes, and dark zones. These all combine to form something like a gigantic vortex, the spirals of which are composed of alternating light and shadow: these extend from the center, a much brighter zone, to fade away at the edges, vanishing in the darkness of a background against which not a single star shines.

We have never seen such a heavenly body from the earth. And, I might add, we shall never see this one from the earth, since it is within this body that the earth is located. All the stars that we see scattered in the sky or in the belt of the Milky Way, all the clusters and nebulae that we have encountered and others so distant that we shall never see them—double stars and variables, novae and supernovae, gas and dust, matter and the radiation in which we are immersed—everything that we have become acquainted with so far combines to form a kind of immense astral city, which includes our solar system as well. The sun is also immersed in it, but if we look for it, we shall not be able to see it. It now appears to be a star of the 19th magnitude (that is, 160,000 times fainter than stars barely visible to the naked eye), and it is lost among tens of millions of stars like itself (Fig. 6.7).

This enormous city is called the Galaxy, and we can now contemplate it at its full extent, as if we were overlooking a panoramic view, because the cluster M 13, in which we find ourselves, is located outside it.

Other than the two splendid, extraordinary visions of the globular cluster and the Galaxy, we shall see practically nothing else from this planet. Apart from some star or another belonging to the cluster itself and close enough to stand out in directions different from that of the cluster or the Galaxy, the sky appears to be completely devoid of stars. And when the planet traverses the intermediate zones of its orbit, we see part of the cluster jutting from one side of the horizon and the Galaxy from the other; the two

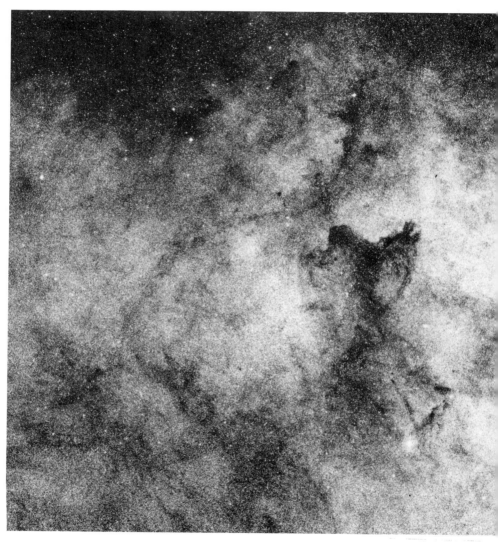

Fig. 6.7   Cloud of stars (in the Sagittarius region), which are so faint and apparently so close together that they form a single milky background. Observed from a hypothetical planet near M 13, the sun would look like one of these small stars, no more distinguishable from the others than a grain of sand on a beach. (*Mount Wilson and Palomar Observatories*)

are separated by a dark, starless chasm. Having nothing similar on Earth, we find it very difficult even to imagine such a sight.

But now let us return to contemplate our galaxy, which is the next area we shall explore. We shall venture into that endless empty space in an attempt to discover whether there is something hidden even there.

7 THE GALAXY

STRUCTURE OF THE GALAXY

The disk we find before us—which seems to extend over the whole sky, almost as if it were the entire universe—is in fact almost inconceivably vast. Thus to form an idea of the dimensions, let us again have recourse to a scale model. Let us take 1 centimeter to represent 10,000,000 kilometers. According to this scale the sun has a diameter of 1.4 millimeters (a shot pellet), the earth is represented by a microscopic dust particle 15 centimeters away, and the closest star, Alpha Centauri, by another pellet at 41 kilometers; beyond these two pellets, at intervals shorter or longer than the first, we find other larger or smaller pellets in groups of two or more, sometimes forming swarms of some hundreds, within which the separation is as little as a few kilometers. On this scale the entire disk extends over a diameter of 926,000 kilometers. A speck of dust near a small pellet in a disk with a diameter greater than twice the distance between the earth and the moon: these are the earth and the sun in the Galaxy.

Returning to the real dimensions, the diameter of the Galaxy is so large that it takes light 100,000 years to traverse it from end to end. The thickness of this disk is considerably less, around 1,700 light-years, although it bulges to 16,000 light-years in its central region.

Although it contains most of the matter, the disk is not the only component of the Galaxy. Around it there extends an almost spherical zone, called a halo or corona, composed of globular clusters, a cloud of scattered stars that do not belong to the clusters, and a small percentage of gas (Fig. 7.1). Both the stars and clusters become more numerous the closer we get to the disk, especially toward the center of the Galaxy. Many of the stars observed in the halo are RR Lyrae–type variables, which, since they belong to the Cepheid category, can be located precisely by means of the period-luminosity law. This also holds for the globular clusters that contain RR Lyrae variables. The distance of the other clusters is derived from estimates of their brightness and diameter. The galactic halo appears to be mostly a combination of stars and clusters at a low concentration, so low that for an observer located at any point in it, not only the stars but also the globular clusters (except for the closer ones) would be invisible to the naked eye. Let us now descend toward the disk, in which the sun itself is contained. Here the number of stars increases markedly, and we find in great abundance everything we are already familiar with: clusters; gaseous nebulae; associations of stars in the process of formation, immersed in clouds of amorphous matter; and, between the stars, great quantities of dust and gas.

Most of these millions of stars and thousands of clusters and associations, as well as the interstellar gas, are not distributed uniformly throughout the whole disk but are primarily gathered in wide spiral arms, separated from one another by relatively empty regions. It was, in fact, very difficult to detect these spiral arms (which we saw so well from the cluster M 13), because the earth, our observatory, is inside the Galaxy and optical exploration at great distances is impeded by the interstellar dust.

It was not until 1951 that W. W. Morgan and other astronomers at Yerkes Observatory were able, by very carefully measuring the distances of bright stars or other relatively bright objects such as clusters and nebulae, to trace a few fragments of three spiral arms near the sun (one of which contains the sun itself). It was thus possible to locate, in the heart of the Galaxy, the Orion nebula, the Cygnus clouds, and other famous regions. This entire area fell within a radius of about 16,000 light-years, which is equivalent, on the galactic scale, to little more than the neighborhood of the sun.

That same year, however, a discovery was made that very shortly would solve even this problem: the discovery of galactic neutral hydrogen. The possibility of distinguishing neutral hydrogen in space by means of a line emitted (or absorbed) at a wavelength of 21.11 centimeters had been proposed as early as 1944 by H. van de Hulst. This line is due to the different spin of the proton and the electron.

As is well-known, hydrogen is composed of a proton and an electron; given that each particle spins on its own axis, if we were to look at them from the same pole, both would be turning in the same direction or else they would be turning in opposite directions. In the first case we say that the spins are parallel and that the atom is at a higher energy level; in the second that they are opposed and the atom is at a lower energy level. The difference in energy between one state and the other corresponds to the emission (or absorption) of radiation of wavelength 21.1 centimeters. This excitation of a hydrogen atom can result from a collision with electrons or other atoms, but since there appears to be scarcely one atom of gas per cubic centimeter in the spiral arms of the Galaxy, this event is very rare, occurring only once every few million years. Moreover, the atom thus excited returns to its fundamental level, emitting 21-centimeter radiation, after about ten million years. Therefore, we would not be able to observe the event in the case of a single atom. But in the depths of space we may come across an enormous number of atoms aligned along our line of sight up to very great distances, since gas (unlike dust) is transparent. We could expect then to pick up possible 21-centimeter radiation.

At this wavelength the radiation is not visible to the eye or with optical instruments; it can be detected only by means of radio telescopes. During the Second World War radio astronomy came under military secrecy acts, but when it was publicized immediately afterward, H. I. Ewen and E. M. Purcell at Harvard confirmed van de Hulst's prediction with the discovery of interstellar neutral hydrogen.

Neutral hydrogen was present mainly in the vicinity of the equatorial plane of the Galaxy. Before long Dutch and Australian radio astronomers began a systematic search in order to locate it exactly and to obtain an almost complete map of it (Fig. 7.2). In the map the spiral arms were clearly delineated (though somewhat tangled), concentric with the nucleus of the Galaxy. Unfortunately, the imprecision of optical measurements and the relatively limited distances they can cover made it difficult to compare the spiral arms defined by the stars with those formed by neutral hydrogen. Thus we still do not know whether the neutral hydrogen arms coincide exactly with the spiral arms defined by the stars and the nebulae of ionized hydrogen (such as the Orion nebula). One thing is certain, at any rate: the structure of the Galaxy is spiral. This is shown not only by the three fragments laboriously traced in 1951 but by all the neutral hydrogen that occurs in a similar pattern over the whole of the equatorial plane.

BETWEEN THE STARS

When we had traveled just beyond the solar system, we thought that only empty space lay between one star and another. Then, after coming upon the nebulae and the dark clouds, we realized that, at least in some regions, this is not so (even though the density of these nebulae is less than that of the highest vacuum obtainable in our laboratories). Now we learn that interstellar space also harbors a lot of dust and gas.

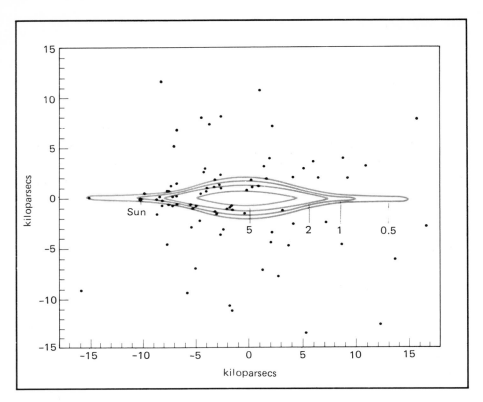

Fig. 7.1   The galaxy and the globular clusters seen in section. The position of the sun is also shown. The curves indicate the density of the disk (taking the density near the sun as 1).

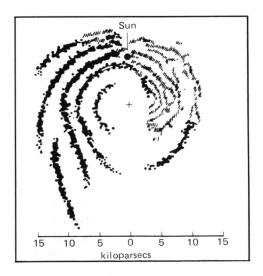

Fig. 7.2   Distribution of neutral hydrogen in the Galaxy based on radio astronomy surveys. The sign + indicates the center of the galaxy.

THE FIRST DISCOVERY

The first element detected was calcium, the lines of which were found by J. Hartmann in 1905, in the spectrum of the star δ Orionis. From the exceptional intensity of the H and K lines in absorption, and by comparing these lines with those present in the spectra of much closer stars of the same type, Hartmann concluded there was calcium interposed between us and the stars, which was responsible for the lines. Later the presence of other elements (such as iron, potassium, and titanium) and molecules (such as the radicals CN and CH) was discovered. The gas is distributed with a density of one atom per cubic centimeter, but in the denser clouds this figure increases at least a thousand times.

Interposed between us and the stars, this gas can be detected as long as the background stars are visible. Unfortunately their light is weakened or even extinguished by interstellar dust, even if the amount of dust, in small volumes, is infinitesimal: it is estimated that in every cubic kilometer of space (again, excepting the denser regions) there are from twenty-five to fifty dust particles no larger than a thousandth of a millimeter across. When we add the absorption effect over great distances, the diminution of the light of the more distant stars becomes considerable. Local, more dense dust clouds at various distances also cause significant light absorption. Furthermore, the stars beyond the more distant dust clouds must be farther, and therefore fainter, on average than the stars beyond the nearer dust clouds. In combination, the preceding effects reduce the contrast between dust clouds and other regions to such an extent that, for distances of over 3,000 light-years, even the dust clouds can no longer be distinguished.

NEUTRAL HYDROGEN

This limitation of the light waves is greatly reduced for longer wavelengths, and there is practically no limitation for radio waves. We took advantage of this property to discover the spiral distribution of neutral hydrogen. We will find it of further use as we attempt to learn more about hydrogen itself and to find out whether other elements or compounds exist. This is a matter of great importance, for although we have seen what elements and compounds make up the heavenly bodies, we have so far been dealing—except in the case of the planets of our solar system—with bodies at very high temperatures. Now we should like to know what we may expect to find in interstellar space, where the temperature is almost down to absolute zero but is basically closer to that of the planets.

Let us start with the neutral hydrogen that we have just encountered. It is distributed over the galactic plane, within the relatively small thickness of 300 light-years, and appears in the form of spirals and arcs. It is also often present in small clouds having diameters of some 20 light-years, masses less than 50 times the mass of the sun, and a mean density of 20 atoms per cubic centimeter. The total mass of neutral hydrogen is thought to represent two percent of the mass of the entire Galaxy.

In June 1970, by means of observations of the ultraviolet spectrum carried out beyond the earth's atmosphere, hydrogen was also found in the molecular state.

THE OH RADICAL

Meanwhile the radio astronomers kept astonishing us with increasingly extraordinary discoveries of other molecules present in interstellar space. The series began with the OH radical. The OH molecule, composed of an oxygen atom and a hydrogen atom, is well-known in chemistry (as hydroxyl) primarily because it is a basic building block for a great many other molecules, including that of water. Atomic physics teaches us that the fundamental energy level of this molecule is divided into four levels and that the electron transitions from one to another can cause the emission or absorption of radiation of frequencies 1612, 1665, 1667, and 1720 megahertz, all of which are near the wavelength 18 centimeters.

The presence of clouds of OH in cosmic space was discovered by chance in 1963 by radio astronomers at the Massachusetts Institute of Technology. The clouds showed up as regions that absorbed radiation coming from intense background radio sources at around 18 centimeters, as anticipated. In 1965 physicists at Berkeley and Harvard also discovered OH emission. Comparison with the distribution of hydrogen soon showed that the concentration of OH increases more than that of neutral hydrogen (H I) toward the central regions of the Galaxy. In fact, in the neighborhood of the sun the proportion of OH to H I is 1:5,000,000; proceeding toward the center, it changes to 1:1,250,000; at about 12,000 light-years from the center it becomes 1:1,000,000; and in the central zone it reaches 1:250,000. That hydrogen and OH do not appear to be mixed, so as to form the same clouds, was shown by the discovery that, at least in some central regions of the Galaxy, they move at different speeds.

In addition to these discoveries about the large-scale distribution of OH, it was found that this molecule often appears in H II zones; that is, it is associated with those nebulae of ionized hydrogen (H II) where we found the associations containing stars in the process of formation. It was suspected at once that these OH regions are not very extensive. The first measurements of their dimensions showed that their apparent diameters should not exceed 5'–10'. With the object of increasing the resolution, interferometric measurements were made, using two radio telescopes 700 meters apart but combined so that they formed one instrument. These measurements, carried out independently at MIT and at the California Institute of Technology, showed that the diameters of the sources must be less than 22".

Subsequent research, carried out by linking more widely dispersed observatories, steadily lowered the value of the maximum possible diameter. It was then decided to combine the parabolic antennae of four observatories in Massachusetts, California, West Virginia, and Sweden. By refining the technique so as to ensure the exact simultaneity of the observations, it was possible to use, for the first time in the history of astronomy, a colossal instrument with a maximum diameter of fully 7,720 kilometers (the distance between the station in California and that in Sweden).

In this way the great discovery was finally made. First of all it was found that one of these sources, located in the area called W 3, was made up of seven nearly pointlike sources. Then, using the Sweden-California baseline, it was possible to measure the apparent diameter of the smallest source; this proved to be barely .005 second of arc. At the distance of W 3 the object turned out to have a diameter equal to that of the orbit of Mars. Thus it could only be a star, gigantic but still having a diameter smaller than that of the largest stars known.

This discovery was important not only because it revealed the presence of mysterious stars whose existence had never been suspected but also— and above all—because of the fact that the energy emitted, being concentrated in a relatively limited volume, must be enormous. In the case of the source in W 3, for example, calculation showed that the star must have a temperature of 10,000,000° K. This temperature is too high to be accounted for by any known mechanism for the emission of energy. Perhaps this emission is produced by a process of maser-like amplification. Assuming that the molecules that produce the lines in the radio spectrum of OH are in some way raised to excited levels by radio waves of a suitable frequency, they will then go to a lower level, relinquishing their energy to an incident wave, which may thereby be greatly amplified. Of course the question of how the molecules of OH are excited remains to be solved. It could happen in various ways: for example, by means of ultraviolet or infrared radiation. The latter radiation is the very type that the protostars, at relatively low temperatures, must predominantly emit. The Russian astrophysicist Shklovsky has in fact put forth the hypothesis that the OH regions coincide exactly with those protostars which precede the phase of nebular variables and which so far we have been unable to observe.

This hypothesis is increasingly supported by other discoveries. In 1967 E. E. Becklin and G. Neugebauer found in Orion a star-like infrared source, which they considered to be a protostar; a while later the radio astronomers E. Raimond and B. Eliasson found that the OH source in emission already noted in Orion coincides exactly with that infrared object. But that is not all. Neugebauer and R. Leighton observed a great number of infrared stars; by consecutively studying twenty that were not associated with H II regions, they found that four are the sites of OH sources in emission.

Thus the OH sources seem to be associated with the protostars—but not only with protostars. More recently various OH sources have also been found to be connected with long-period variable stars. Hence there would be OH masers of two types: one associated with young stars and the other with stars in a very advanced stage of evolution. In both instances we are dealing with stars that have thin but very extensive atmospheres; in the first case, these atmospheres appear very frequently in H II zones—that is, in those clouds of ionized hydrogen with which they have been associated from the start.

OTHER MOLECULES

The discoveries concerning the OH radical should not cause us to neglect the other molecules. Discoveries are being made in this area that are extremely promising and very baffling, to say the least.

In 1968 ammonia (NH_3) was found in emission. Early in 1969 water vapor (H_2O) was found in the Orion nebula, in the Sagittarius B 2 cloud, and in the W 49 cloud in Aquila. We have already encountered water vapor in the outer layers of certain stellar atmospheres; now by the existence of actual interstellar clouds of vapor, we have a new confirmation of its presence and abundance in the universe. In March 1969 the first polyatomic organic molecule was discovered in space, that of formaldehyde (H_2CO). Unlike NH_3 and H_2O, which are present only in certain regions, H_2CO was found in various parts of the Galaxy—in fact, in 60% of the sources where it was sought. It was found that in the central zones of the Galaxy H_2CO moves at the same speed as OH, and this means that OH and formaldehyde may join to form interstellar clouds.

In April 1970 radio waves revealed carbon monoxide (CO) and cyanogen (CN). The latter had already been found optically. The former is particularly important because it is present in formaldehyde. It is not yet quite clear, however, whether it must be present to form formaldehyde by combining with the very abundant interstellar hydrogen, or whether it results from a primitive formaldehyde being split into H_2 and CO by intense radiation.

In June 1970 hydrocyanic acid (HCN) was found; it is of great interest organically because, at least in our laboratories, it plays an important role in the formation of the simpler amino acids. A month later the molecule of isocyanic acid (HC_3N) was found in emission in the radio source at the center of the Galaxy, and in November the molecule that was probably least expected was found: methyl alcohol (CH_3OH), observed in the direction of the two sources Sagittarius A and Sagittarius B.

The discovery of interstellar molecules continued at an increasing rate. In January 1971 the discovery of formic acid (H_2CO_2) was announced. It is the simplest organic acid, present on Earth in nettles, bees, and ants, and was found in the Galaxy in emission in the radio source Sagittarius A. In March formamide (NH_2COH) was detected. From then until the end of 1975 more than twenty new molecules were discovered, mostly in the radio sources at the center of the Galaxy. One of these, methylamine (CH_3NH_2), discovered early in 1974, is particularly important; it can react with formic acid to form glycine, the simplest of the amino acids. The proteins, which together with nucleic acids constitute the principal building blocks of life on our planet, are formed from these acids. February 1975 saw the discovery of acrylonitrile (CH$_2$CHCN), which is used to make synthetic fibers or rubber and certain plastic materials. And the search and discoveries are continuing.

We cannot dwell on some of the points that are being studied even now, such as the dimensions of the clouds, their origin, the variability of the radio emission. But one fundamental fact has now been established in the last six years: there are a great number of organic molecules in interstellar space. When we tried to discover the composition of the heavenly bodies, we found there the same elements that are present on Earth, but since we were studying bodies at high temperatures, we found only atoms, and even these were often ionized. As soon as we considered stars of lower temperatures, we found molecular compounds. And now, having scarcely begun to explore interstellar space with means to suit the task, we discover that the immense, cold void contains molecules that are actually organic or, like H$_2$O and NH$_3$, closely linked with our organic world.

That is not all. We have found that the majority of these molecules are present only in relatively limited regions, where many stars are contemporaneously in the first phases of evolution. Probably it is just that brief period in which a star is born, during the contraction of the primitive cloud, when the molecules appear in that same cloud as it condenses. These molecules must also abound in the atmospheres of the planets, which, as now seems certain, are formed at the same time. Later the remains of the interplanetary gas are blown away by the central star, which, having become much hotter and brighter, dissociates the molecules of which the gas is composed.

Nevertheless some extraterrestrial organic material must remain in interplanetary space. That fact seems to be proved by another great recent discovery: the presence of abiological amino acids in a meteorite.

The recent discoveries of both interstellar organic molecules and amino acids in space have not yet been fully collated, but they already stand out clearly as two fundamental pillars on which we can build a bridge connecting the primitive, fiery world of the stars and cosmic matter with the delicate, sensitive world of living beings. Thus our study of the structure and composition of the Galaxy has led us, perhaps, to the origins of the living world: the factory where the first organic molecules are produced—molecules from which this entire world is then built.

THE GALAXY MOVES

While we were wandering here and there in the Galaxy—first discovering the halo (from a point in which, near M 13, we had set out), then descending to the disk, where we explored interstellar space—we were unable to notice a most important phenomenon: that the Galaxy itself is in motion.

We have already seen that all the heavenly bodies we have encountered are moving: planets, stars, clusters. Now we are about to learn how the entire stellar world that we have just discovered moves as a whole. Do the stars migrate all together, like an immense swarm, in one direction, toward some unknown common goal? Or do they move individually or in little groups, in every direction, like the molecules of a liquid being heated? They do neither: everything turns about the center of the Galaxy.

Yet it is a rather strange rotation. It is not like that of a rigid body (for example, a phonograph record), in which the speed increases with the distance from the center, as would happen if all the bodies belonging to the Galaxy (and hence their masses) were uniformly distributed over the disk. It is not even a Keplerian motion, like that of the planets around the sun, in which the speed gradually diminishes as one moves outward, as would happen if the entire mass of the Galaxy were concentrated in the center (Fig. 7.3). The actual motion is a composite of the two effects (Fig. 7.4). From the center to a distance of about 5,000 light-years the rotation takes place as for a rigid body; between 10,000 and 25,000 light-years the speed of circular rotation changes slightly, corresponding neither to the simple motion of a solid body nor to a Keplerian motion; beyond this limit the speed falls off sharply, and the motion follows Kepler's third law.

This means that within a radius of 5,000 light-years the mass is distributed nearly uniformly and that beyond 25,000 light-years there remains so little mass that this part of the system behaves as if the whole mass were concentrated in the center. In other words, the way the Galaxy rotates indicates that most of the mass—and hence most of the bodies of which it is composed—must be in the central zones.

These findings were obtained at the cost of considerable efforts, both observational and theoretical, over almost a century of research, which has led to other interesting discoveries as well.

One of the first points to be revealed was the position and the motion of our solar system—or, rather, as we shall say henceforth, of its center, the

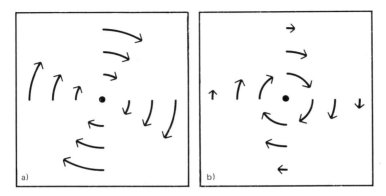

Fig. 7.3   Galactic rotation: *left,* that of a rigid body; *right,* Keplerian motion.

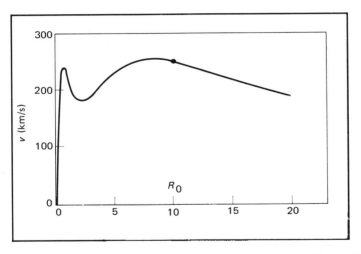

Fig. 7.4   Curve showing the rotation of the Galaxy: it is the resultant of the two types of rotation illustrated in Fig. 7.3.

sun—relative to the Galaxy. According to the most recent estimates the sun is 30,000 light-years from the center of the Galaxy. We can measure its motion with reference, for example, to the globular clusters, which constitute a system that is outside the disk and certainly not subject to its motion. It turns out that the sun moves in the galactic plane, in a direction perpendicular to that of the center—that is, in an almost circular orbit, which it traverses clockwise (as seen from the north galactic pole) at a speed of 250 kilometers per second. This is indeed an enormous speed; it impresses us all the more since it is not that of some remote body studied by astronomers but the one at which, in this very moment, we ourselves are moving, as we follow the sun along its course in space. When we boast of having taken five minutes less to drive from Rome to Florence, let us not forget that, carried by the Earth, we have covered the same distance in only one second and that any sedentary person can, without stirring from his armchair, boast of having covered tens of millions of kilometers by the end of each day.

Proceeding at such a pace, we might expect to be able to make a quick tour of the Galaxy and, transported by this magnificent spaceship called Earth, see a rapidly changing array of new panoramas.

Unfortunately the matter is not so simple, for two reasons. The first is that, although we do make a complete circuit of the Galaxy, it takes a long time, despite the extremely high velocity, because we have such a long way to go. Let us not forget that the radius of the circle that we cover is 30,000 light-years. Thus the entire route is 188,400 light-years, which, even at 250 kilometers per second, we can cover only in 226 million years. This figure seems enormous, and indeed it is, but only relative to the shortness of our life and the life of the species that inhabit the earth.

When the first men appeared, a million years ago, the sun and the planets, including the earth, were essentially in the same region where we find ourselves today: that is, scarcely 1 centimeter back along a route measuring 2.26 meters. If, however, we go back in time through the geological eras, we can connect various events that occurred on the earth's surface to the displacement of the earth itself within the Galaxy; this should provide us with very expressive terms in which to make comparisons in time and space. Thus we note that the last saurians, on the way to extinction, lived in that part of the Galaxy diametrically opposite the part in which we now find ourselves, and their progenitors in the region immediately preceding it. The first mammals, on the other hand, appeared in the same position we are in now, but exactly one revolution earlier: thus in one complete revolution of the Galaxy the mammals developed from the first primitive types to their present states. Two revolutions ago—again, when the earth was in the same position—the first reptiles, amphibians, and vertebrates appeared. From the time when life began to develop in the most rudimentary forms and through the first invertebrates, our planet has completed at least three revolutions about the center of the Galaxy. This is still only a small part of its voyage in space, for, since the time of its formation, or rather the formation of its solid crust (about 4.5 billion years ago), it has made this long circuit some twenty times; man, from his origins until now, has covered only 1/226 of that journey.

The second reason we would not see such a rapidly changing view is that the sun is not moving among fixed stars, for, as we have just seen, the entire disk of the Galaxy is in rapid rotation. On the other hand, the disk does not move as a rigid body, at least near the sun. Thus the stars, depending on whether they move more rapidly or slowly than the sun, show a relative motion that makes them shift with respect to us, but much more gradually than would be the case if only the sun were moving among other stars fixed in their positions. Let us consider the stars near the sun (Fig. 7.5): they all move along paths parallel to the sun's, at approximately the same velocity—a little faster than those closer to the center of the Galaxy and a little slower than those farther away. The observed velocities, which are simply the differences between the sun's velocity and the velocities of the respective stars, are generally less than 40–50 kilometers per second. These stars are called low-velocity stars. A similar behavior is seen in the interstellar matter and the bright nebulae, indicating that they also follow the motion of the solar system in space.

There are, however, a great many stars that appear to move at high velocities, over 80 kilometers per second. These stars, known as high-velocity stars, do not follow trajectories parallel to the sun's path but move in highly eccentric orbits that intersect the sun's orbit (Fig. 7.6). Hence they are close to the sun only temporarily. Their actual velocities may well be equal to that of the sun, but their relative velocities appear different because

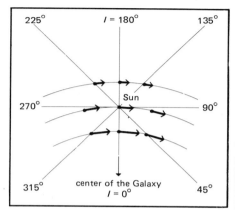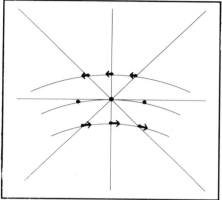

Fig. 7.5 Differential rotation of the Galaxy: *left,* velocities of the stars relative to the center of the Galaxy as they would appear to an outside observer; *right,* how the motions of the same stars appear as seen from the sun.

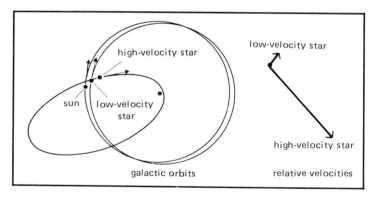

Fig. 7.6 Stars of high and low velocity: *left,* the orbit of a high-velocity star lies in a plane that is outside the galactic plane, whereas the orbit of a low-velocity star, such as the sun, lies within the galactic plane; *right* (on a larger scale), the velocities relative to the sun of a low-velocity and a high-velocity star.

the directions of the respective motions are different. Moreover, whereas the magnitude of the sun's velocity is nearly constant, since its orbit is circular, that of such stars, given the considerable ellipticity of their orbits, varies appreciably from one point to another (according to Kepler's second law), reaching a maximum when close to the galactic plane and a minimum at the greatest distance from it.

The orbits of the high-velocity stars do not all lie in the galactic disk; rather, they lie in planes inclined at every possible angle to the plane of the galactic equator, the way comets circulate in the solar system. Thus the high-velocity stars move away from the central regions of the Galaxy and make periodic excursions to the halo. Better yet: they are actually halo stars that dive into the disk only for a part of their orbit. The globular clusters are also thought to follow similar orbits. This would explain why the number of halo stars and globular clusters increases as we approach the central regions of the Galaxy. In fact all the halo stars and clusters must pass near these regions, whereas at great distances from the center, they are scattered in every direction because of the various inclinations of their orbits.

At this point we must note one final important fact. It was shown that the high-velocity stars belong to Population II, and since the globular clusters are also made up of stars of this type, we must conclude that the halo is composed exclusively of Population II stars.

Thus the disk and the halo not only show different structural and dynamic characteristics, they are composed of two different types of stars: Population I and Population II. Recalling that these two populations correspond to different ages, we thus discover that the present Galaxy, as we have come to know it, was formed in two different epochs: a more remote epoch in which the halo stars were born, and a later epoch in which the disk, as well as all the material that composes it, was formed. And in the disk the birth of Population I stars is still going on.

THE MASS OF THE GALAXY

We have discovered that all the stars and bodies we have seen so far, along with a great many others that we are certain exist, even though we have never been able to see them, form an immense system, so vast that it takes light (which travels from the earth to the moon in little over a second) a good hundred thousand years to cross it. We have explored the boundaries of this system and glimpsed the void beyond, we have traversed it from the halo to the disk, and we have discovered how the stars move within it. Yet we still do not know how many stars it contains; like the astronomers of past centuries, we have no idea how many stars we could count (if we had the time and the means) between here and the point where they end, that is, from here to the limits of the Galaxy. Unfortunately we shall never be able to determine the exact number, not only because it would be difficult to construct instruments that would enable us to perceive even the most distant faint stars but above all because a large proportion of them are hidden by the dust clouds in which we are immersed. Nevertheless we can still make a fairly good estimate of the number of stars in the Galaxy by an indirect but efficacious method: weighing it as a whole.

The weight, or rather the mass, of the Galaxy is not difficult to determine once we know its motion. We have seen that all those stars which are sufficiently peripheral to permit the assumption that most of the total galactic mass is contained in the center of the Galaxy move in Keplerian orbits. Among these is the sun. Thus applying Kepler's third law (in Newton's version, including the mass) and disregarding the mass of the sun in comparison with that of the Galaxy, we have:

$$M_{Galaxy} = \frac{R^3}{P^2},$$

in which the distance R of the sun from the galactic center is 2 billion astronomical units and the period of revolution P of the sun is 200 million years. We then have:

$$M_{Galaxy} = \frac{(2 \times 10^9)^3}{(2 \times 10^8)^2} = \frac{8 \times 10^{27}}{4 \times 10^{16}} = 2 \times 10^{11} M_{Sun}.$$

This, then, is the total mass of the Galaxy: 2×10^{11} solar masses, or 200,000 million times the mass of the sun. Assuming that at least half of this mass is due to normal stars and the rest to dust, gas, and dark bodies (such

Fig. 7.7   The Sagittarius region of the Milky Way in a wide-field photograph taken with a lens of focal length 25 cm. The circle indicates the center of the Galaxy.

as planets), and that the average mass of the stars is equal to that of the sun, we find that the Galaxy contains at least 100,000 million stars.[1]

Now the picture is complete, and we finally have a clear, quantitative idea of the sidereal world, which for centuries was thought to be boundless: it is made up of billions of stars and a corresponding mass dispersed in gas and dust, constituting an immense disk and an extended halo. The disk revolves about its center; it also represents the focus of the elliptical orbits of the halo's stars and clusters, which plunge near it periodically. This point, then, is not only the geometrical center of the entire system but also the only fixed point in the entire sidereal world; everything we have seen so far—stars, clusters and associations, gas and dust, the earth, the sun, ourselves—revolves around it. And indeed, the region surrounding this point is the only part of the Galaxy we have not yet explored; its exceptional position gives us good reason to believe that we might find there something different from anything we have seen so far.

So let us set forth on this new journey, which will bring us to the center of the vortex, where we should find an exciting spectacle of multitudes of stars moving around us, unless something we find there is so extraordinary and so much more important that it diverts our eyes and thoughts toward new, even deeper mysteries.

THE CENTER OF THE GALAXY

As we look out from the earth toward the center of the Galaxy, the first thing that strikes us is an enormous cloud of stars that seems to stand out clearly from the rest of the Milky Way, which serves as a backdrop (Fig. 7.7). The effect is not very evident from the latitude of Italy, where the area always appears low on the horizon, even on the most favorable summer evenings, whereas from the Southern Hemisphere it is seen in all its grandeur. This cloud of stars does not, however, correspond exactly to the center of the

[1] In reality this calculation is valid only as a rough approximation, since, gravitationally, the Galaxy does not behave like a point source. The distribution of the mass among the various bodies is derived from local examination, which, besides showing that about half of the mass belongs to the stars, reveals the existence of a considerable hidden mass.

Galaxy, which lies slightly to the right in a region that appears less dense because of the presence of dust. The cloud of stars and, above all, the dust clouds prevent us from seeing to any great depth; thus what we manage to see, whether with the naked eye or in normal telescopic photographs, is not the galactic center, but rather the barrier, more or less opaque, that hides it. This barrier, of varying thickness, is already relatively close to the center and thus constitutes, in a certain sense, its outermost regions, which could themselves provide us with important information; however, we now propose to pass through these regions in order to discover what is hidden behind them.

As we gradually approach the central regions of the Galaxy, it becomes harder and harder to see, and in presuming to distinguish directly the matter of which these zones are composed and the phenomena that occur there, we find ourselves increasingly in the position of one who, wishing to understand the structure of a volcano, descends into it. Even if an excursion such as the one we are imagining were to leave us unharmed, we should not gain much by it since, enveloped in smoke and flaming gases, we should see nothing, and however close we came, an impenetrable curtain would prevent us from seeing the source of those impressive phenomena that we can see so well from outside the central regions and at a certain distance. Here not even theory can help us (as it did for the interior of the sun); very little is known about the masses, densities, and temperatures, and still less is known about the motions of the bodies found within the regions closer and closer to the center.

Nevertheless we can still use means other than the naked eye or photography to gather data that, when properly collated and interpreted, may also provide a meaningful image. Inside a cyclone we can see nothing, but by measuring the direction and velocity of the wind, the precipitation, temperature, atmospheric pressure, and the variation of these quantities with the passage of time as the phenomenon develops, we can construct a fairly good overall picture of it. Moreover, recent years have seen the development of methods particularly suited for observation of the central regions of the Galaxy: radio astronomy and infrared astronomy. With the former we can chiefly study the distribution and behavior of the gases (hydrogen, the OH radical, etc.); with the latter we can reach regions much closer to the nucleus by penetrating as deeply as possible the dust clouds that envelop it. It is as if, descending into the volcano, we had recourse to one method for studying the vapors and another for penetrating as far as the fire, and thus were able to lessen—though not eliminate completely—the great impediment of the smoke.

We come across the first surprise at about 12,000 light-years from the center of the Galaxy. Here the radio telescopes have shown two spiral arms of neutral hydrogen (Fig. 7.8), one between the sun and the galactic center, and another beyond the center. The first, observed for about a quarter of its circumference, is estimated to have a total mass twenty million times that of the sun. The presence of these arms seems in no way extraordinary; we have come across so many arms of neutral hydrogen throughout the Galaxy that the presence of these two segments in a region close to the center does not alter the nature of our knowledge. But these two arms have shown a characteristic as extraordinary as it is unexpected: they are expanding rapidly from the center. The one toward the sun is moving out at a velocity of 53 kilometers per second; the other at 135 kilometers per second. Subsequent observations have shown that there are also clouds and extensive regions of neutral hydrogen moving rapidly away from the galactic center in directions inclined to the equatorial plane, in which the first two arms lie. In short it appears that most (perhaps 70%) of the neutral hydrogen in this region of 12,000 light-years' radius—a total mass estimated to be equal to at least 50 million solar masses—is fleeing from the central regions of the Galaxy at a speed of 53 kilometers per second or more.

This expansion is very mysterious. Perhaps it is the result of an explosion that took place in the center of the Galaxy some millions of years ago—for what reasons, we do not know.

Inside the two arms is a disk of neutral hydrogen (Fig. 7.9), 5,000 light-years in diameter and with a mass four million times that of the sun; it is in rapid rotation, with a speed of 230 kilometers per second at the edge and 200 kilometers per second at a distance of 1,200 light-years from the center. It is thinner in the central region, where it runs about 230 light-years thick; it grows thicker toward the edge, however, finally stabilizing at about 800 light years, which is the thickness within which neutral hydrogen is contained along the entire equatorial plane of the Galaxy.

In addition to hydrogen, the central region of the Galaxy contains abundant clouds composed of other matter. Hydroxyl, as we have already seen,

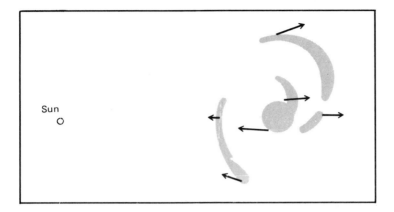

Fig. 7.8   Central zone of the Galaxy, where Dutch radio astronomers have discovered parts of two spiral arms expanding from the center.

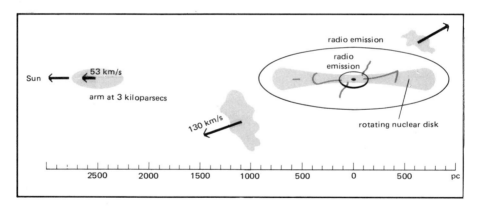

Fig. 7.9   Enlarged detail of the central zone of the Galaxy along a section perpendicular to the galactic plane: one can see the spiral arm 3 kpc (kiloparsecs) from the center and two hydrogen clouds moving away from the center in opposite directions, outside the equatorial plane of the Galaxy. The small black ellipse at the center is extremely abundant in gas, dust, and stars, and emits radio waves, infrared radiation, and X rays.

is extremely plentiful, and, from all appearances, the clouds it forms do not coincide with those of hydrogen. Radio observations show that the two types of clouds move at different velocities. Thus while it appears that the hydrogen only rotates around the center of the Galaxy or streams away from it, some clouds of OH are actually falling toward the center. Let us recall, moreover, that the hydroxyl molecule is not the only molecule present in this region: recently ammonia, water vapor, formaldehyde, and many other molecules have been found there. Unfortunately little is known as yet about the structures, positions, and motions of these clouds; thus, while not forgetting that they too contribute to the overall picture, we shall not study them in detail as we examine the galactic center further.

The disk of hydrogen we encountered a while back is mostly contained within the central bulge of the Galaxy, which has a diameter of 16,000 light-years. This bulge is very rich in stars, the more numerous the farther one proceeds inward. In the innermost part, to a diameter of 6,500 light-years, is contained one-tenth of the mass of the entire Galaxy. The stars seem to be mainly Population II stars, but not entirely; many type-M supergiants have been observed. There are globular clusters, but they are smaller than those we encountered in the halo and their stars are more concentrated. The number of stars increases rapidly toward the center, but unfortunately the proofs we have of their presence are increasingly indirect since observations are increasingly obstructed by dust. At 330 light-years from the center the stars are already a thousand times as numerous as in the sun's neighborhood; at 33 light-years sixty thousand times; and at 3 light-years half a million times. Thus the stars in the central zone of the Galaxy increase much more rapidly than do those in globular clusters, and they are a thousand times more densely packed.

While the number of the stars increases at a dizzying pace, the density of the hydrogen does not seem to vary much. All the same, the mass of neutral hydrogen within a radius of 330 light-years is still equal to a million times the solar mass.

But in approaching the nucleus we do not expect to find only stars and gas clouds, as elsewhere in the Galaxy, even if in different proportions. Even before we knew anything about it, we felt that the nucleus must be different from everything else, if only because it is the only point that remains fixed while everything moves around it. Moreover, the expulsion of gas that we have already seen, the great percentage of mass concentrated near the nucleus, and its stellar population (different from that of the disk and instead strangely like that of the halo) all give us the feeling that it contains something unusual, perhaps the key to the mystery of the very existence of the Galaxy. That it is exceptional has been confirmed by a recent discovery: an emission of X rays, fairly extensive if not very intense, observed in the first months of 1971 by means of the artificial satellite *Uhuru.* The origin of the emission remains a mystery. Therefore, let us tackle the exploration of the restricted innermost region of this very important kernel that is at the center of everything.

First of all we find that the center of the Galaxy contains an intense radio source. It was discovered in 1951 and was called Sagittarius A. In 1955, however, F. Drake succeeded in resolving it into five distinct sources, and the name Sagittarius A has since been reserved for the most intense source. The radio emission comes from a region 30 light-years in diameter, but the most intense region, Sagittarius A, is much smaller and completely encloses the dynamic center of the Galaxy.

Since 1968 the center of the Galaxy has been studied much more closely, especially in the infrared.[2] Observations made in the interval between 1.5 and 4 micrometers have shown an intense infrared source with a

---

[2] W. Herschel discovered infrared radiation (that is, radiation at wavelengths longer than that of red light) early in the last century, and since then various heavenly bodies have been observed repeatedly in this spectral region. However, research and discoveries in the infrared have developed substantially only in the last fifteen years or so. By infrared we actually mean radiation of wavelengths between 0.75 and 1,000 μm (micrometers). Below this is visible radiation (light), and above, radio waves. It is not possible to explore all of this interval from the ground. For example, the radiation coming from celestial bodies between 25 and 1,000 μm is blocked by the earth's atmosphere, which intercepts it or hides it with an emission of its own. The interval from 0.75 to 1.2 μm can be observed photographically. Between 1.2 and 25 μm there is already atmospheric absorption, but with a few exceptions in certain regions called "windows." (The chief windows are at around 1.6, 2.2, 3.4, 5.0, 10, and 22 μm.) Observations in the infrared beyond 1.2 μm are carried out by attaching suitable detectors to the telescopes. If one then wishes to make observations in regions outside the windows, the entire instrumentation must be raised to great heights, using aerial means (airplanes or balloons); if it is necessary to go outside the atmosphere, the instrumentation must be mounted on rockets or artifical satellites.

diameter of less than 15 light-years that coincides perfectly with Sagittarius A (that is, with the nucleus). Much of this radiation is almost certainly emitted by red stars or stars greatly reddened by the enormous quantity of dust present in the neighborhood. If this is true, in the most central part of this region, scarcely 3.3 light-years in diameter (a little less than the distance between the sun and the nearest star), there would be ten million times as many stars as there are in the sun's neighborhood. In 1969 a very intense infrared source at wavelengths between 5 and 25 micrometers, barely 2.5 light-years in diameter, was discovered within this region. Its brightness is a million times that of the sun, and we do not yet know what causes it. It may be that we are dealing with a very dense dust cloud, containing in its center an extremely intense ultraviolet source; in that case the dust would absorb the ultraviolet and luminous radiation coming from the interior and would reemit it at longer wavelengths, that is, in the infrared. The ultraviolet source could be due to stars of the earliest spectral types or to nonthermic radiation produced by unknown phenomena.

This is still not the nucleus of the Galaxy. Infrared observations at 1.5–4 micrometers have shown an emission coming from a space that is even more restricted. At the center of the region of 15 light-years in diameter there is an intense source 100,000 times brighter than the sun. It appears to have a diameter no greater than 170 times that of our solar system and a temperature of about 2,000° K. This is not much on the galactic scale; indeed in this sense it would seem to be insignificant in terms of its size and its temperature, which is barely equal to the surface temperature of cooler stars. But in reality it is a highly exceptional body, unique in the entire Galaxy, for no other star has the same characteristics. And, though the temperature is low for a star, it is still that of a fiery furnace, not the size of the earth or the sun or the solar system, but an immense sphere with a diameter 170 times that of the solar system.

This, then, is what lies at the center of the Galaxy: an enormous furnace, a huge crucible. If we penetrate its interior, we shall see nothing, for the dust prevents us from perceiving even the innermost stars of the Galaxy. We shall see nothing except, perhaps, some red star in the nucleus itself, bright enough and close enough so as not to be completely obscured. Let us pause for a moment in this fiery darkness and, while we are still uncertain whether this is indeed all there is in the galactic nucleus or whether there is something else the observations have not yet revealed and which eludes us, let us recall an experiment we made as children. After eating a peach, we came to the pit as always, but this time we did not throw it away. We broke it open in order to discover how, when it is planted, it manages to become a new peach tree. But we found nothing special: only a kind of almond, good to eat (we had not yet been told that it might contain a very dangerous poison). And yet that very seed, which did not look like anything special, must have contained the mystery of life.

Now, at the center of the Galaxy, having experienced the same disillusionment, we feel the same way. We have discovered an immense furnace but a new mystery as well. For we do not know how that furnace came to be here, nor what it is burning, nor how. If we only knew, we might perhaps have solved not only the mystery of the structure of the Galaxy, but also the mystery of its origin and evolution, and perhaps even the origin and evolution of the universe.

THE MAGELLANIC CLOUDS

When, starting from the moon, we began this extraordinary journey, our least concern was the means of transportation that we used. We were well aware that we would need some means capable of penetrating the interior of the stars, of hovering over the most disparate panoramas of the sidereal world, of transporting ourselves at speeds much greater than that of light—in short, a means that is unrealizable, even conceptually. Hence, we relied on the most efficient, versatile, dependable, and economical means: the imagination. Thus traveling from star to star, we reached one of the closest globular clusters, from which we viewed and studied the Galaxy that we had left and in which we then reimmersed ourselves.

Now, leaving the center, where we saw hardly anything, we emerge in a zone of the disk and find ourselves at some point or other, suspended in space and surrounded by an extremely full starry sky; crossed by the band of the Milky Way (itself blazing with stars), which seems to weld the two hemispheres together.

Perhaps we are stationary, immobile in space; perhaps we are moving through it at a speed close to that of light. We do not know, and we make no effort to find out. We are inside a strange sky, surrounded by unknown stars or by stars that, when observed from Earth, appeared to be of quite a different brightness and were distributed quite differently in the celestial vault. But amid the great variety of new objects there is still something that we can recognize, having already seen it from Earth: two irregular milky spots that appear to be detached parts of the Milky Way.

They are the Magellanic Clouds, named after the great Portuguese navigator who first noted their presence in the southern sky, during the first voyage around the world. When we observe them from closer up (or as we are able to do on Earth by means of large telescopes), the two clouds are resolved into a myriad of stars, most of them faint or even extremely faint. Among these, starting early in this century, many pulsating variable stars were discovered that belonged to the class of Cepheids, which we have already encountered. As we saw then, these stars have the strange property of a mean intrinsic brightness that is greater the longer the period of light fluctuation. H. S. Leavitt discovered this law in 1910 while studying the Cepheids in the Magellanic Clouds. Thus by calibrating the period-luminosity

Fig. 8.1   The Small Magellanic Cloud, to the right of which is the globular cluster 47 Tucanae. The latter appears close to the cloud on account of perspective, whereas in reality it is much closer to the earth. (*Harvard College Observatory*)

Fig. 8.2   The Large Magellanic Cloud, which according to the most recent measurements is closer to the earth than the Small Cloud and has a diameter of 35,000 light-years. (*Harvard College Observatory*)

relation among certain nearby Cepheids of known distance, it was enough to know the periods of the light variations to derive the absolute magnitudes of the Cepheids in the Magellanic Clouds, and since the apparent magnitudes were already well-known, it was possible to derive their distance.

The matter later proved more complicated, with the discovery of two types of Cepheid variables, each adhering to a different period-luminosity relation. It was necessary, therefore, to determine the two ratios and to know, in each case, of what type the Cepheids in question were, so as to apply one or the other ratio. At any rate, by refining this method it has been possible to determine very accurately the distance of the Cepheids in the Magellanic Clouds and hence that of the clouds themselves. According to the most recent measurements the Small Cloud (Fig. 8.1) is farther from the earth, at a distance of 200,000 light-years, while the Large Cloud (Fig. 8.2) is closer, at a distance of 170,000 light-years. Thus we are dealing with extremely distant objects, undoubtedly outside the Galaxy. Given the enormous distance and in view of their considerable apparent extent (the apparent diameter of the Large Cloud, as seen from the earth, is twenty-four times that of the moon; that of the Small Cloud eight times), it is at once clear that their dimensions must be enormous.

The Large Cloud turns out to have an actual diameter of 35,000 light-years and the Small Cloud 14,000. Thus they are two true galaxies, outside our own and similar to it, even if their dimensions are considerably less.

Let us examine one of the two in greater detail—for example, the Large Cloud. More than two thousand variable stars have been discovered there so far, of which over six hundred are Cepheids that have been thoroughly studied. Among the many others yet to be studied, some tens of long-period variables, various irregular variables, and many eclipsing variables have already been distinguished. The appearance of five novae, at different times, has also been recorded.

Many stars in the Magellanic Clouds are gathered in clusters. As a result of research conducted by P. W. Hodge and J. A. Sexton in 1966, the total number of clusters discovered and cataloged in the Large Cloud has reached 1,603, but obviously we do not see all that exist; these authors estimate that the total number must be about 6,000. For the most part they are open clusters. However, thirty-five globular clusters are also known, made up primarily of red stars, like those present in the Galaxy. Furthermore, some strange clusters, of a type not observed in the Galaxy, have also been discovered in the Large Cloud. These are stellar aggregations similar to the globular clusters, but their brightest stars are blue, not red. Their existence is, at the moment, somewhat mysterious, but it is thought to be interpretable in the framework of stellar evolution.

Over four hundred emission nebulae have been observed. The largest is an object the like of which is not known in our entire Galaxy. It is an immense nebula, called the Tarantula (Fig. 8.3) because of its shape, having a diameter of 800 light-years and a mass equal to five million times that of the sun. The Orion nebula, though it is one of the most beautiful and largest nebulae in the Galaxy, has a diameter of scarcely 25 light-years and a mass one hundred times that of the sun. If the Tarantula nebula were located at the distance of the Orion nebula, it would cover more than a quarter of our sky, which it would dominate strikingly, with a total brightness about three times that of Venus.

Photographed with suitable filters, the Tarantula nebula was found to be excited not by a single star, but by a cluster, situated within it and composed of blue or white supergiants, like those in the Orion Trapezium—that is, extremely young stars. Thus the Tarantula nebula, like the Orion nebula, is a star factory, but much vaster and fuller (Fig. 8.4).

The wonders of the Large Cloud have not yet been exhausted. We encounter there, in fact, some of the most brilliant stars known (excluding, naturally, supernovae in the explosive stage). The brightest of all is the star R 76, which, having apparent magnitude 9, can be seen even with a modest telescope. Its absolute magnitude turns out to be −9.1. This means that it is nearly half a million times brighter than our sun. If such a star were placed at the center of our solar system, it would immediately blind us. At the standard distance of 32.6 light-years used to determine absolute magnitudes, where the sun would appear as one of the faintest stars visible to the naked eye, R 76 would be bright enough to make objects cast shadows—nearly as bright as the half-moon.

If the Large Magellanic Cloud were closer to us or if we could observe it with more powerful telescopes, we should certainly discover other marvels. But we would have difficulty discovering anything more extraordinary than what we found to begin with: that it and the Small Cloud are two galaxies similar to our own—much smaller, but galaxies all the same, that is, aggre-

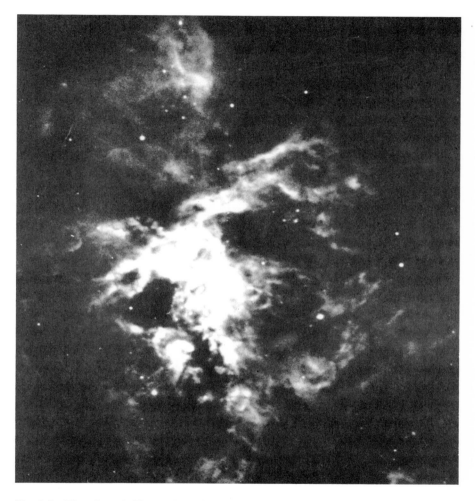

Fig. 8.3 The gigantic Tarantula nebula, 800 light-years in diameter, which is located in the Large Magellanic Cloud, photographed by Westerlund in the spectral line due to the emission of doubly ionized oxygen. (*B. E. Westerlund*)

Fig. 8.4   The central zone of Fig. 8.3 photographed by Westerlund in a region of the continuous spectrum adjacent to the spectral line of the preceding photograph. This device made it possible to exclude the nebula and to see within it a cluster of bright early type stars similar to those of the Trapezium in Orion. (*B. E. Westerlund*)

gations with masses billions of times greater than that of the sun, in which matter is present in the form of stars, gas, dust, and, surely, billions of planets that perhaps we shall never see but the existence of which reason forces us to assume.

Thus continues the game of Chinese boxes. Beyond the earth, which the ancients considered the center of the universe, other earths—the planets—have been discovered. Beyond the sun other suns—the stars—have been found. And now, beyond the Galaxy, much farther than the zone where the stars of our system end, after crossing a vacuum of 150,000 light-years, there appear two new galaxies, composed in turn (like our own) of other suns, other earths. These galaxies are nowhere near as big and rather different in appearance, but they are, all the same, aggregates of stars, gas, and dust, just as our galaxy is. This discovery arouses our curiosity and imagination and inspires us to attempt to proceed farther, beyond the Galaxy and the Magellanic Clouds, in order to see whether the stellar universe really comes to an end and the Magellanic Clouds are merely two exceptional diminished replicas of the Galaxy, or whether we shall find something more—and, if so, to see whether what we find is still similar to what we already know, or whether it will turn out to be completely new, different, and unexpected.

LEAVING THE MAGELLANIC CLOUDS AND THE GALAXY

THE COSMIC VACUUM

Thus let us leave the Magellanic Clouds and proceed further. When we have gone far enough to leave behind us all the stars of the Galaxy and of the two clouds, we shall be confronted with an impressive sight, difficult even to imagine. We are struck, not by what we see, but by what we do not see. We can no longer discern ahead of us a single heavenly body. Our bewildered gaze finds nothing to focus on and is lost in the void. Space, deprived of any object, become a dark abyss that seems to move toward us only to engulf us in nothingness. There is nothing in our daily experience that could suggest anything comparable to this sight. In the darkest night we still glimpse something; when we wake up in the complete darkness of our bedrooms our eyes see nothing, but we do not have the sensation of an enormous void, for we still sense the presence of the objects that are around us, and even if we do not see them, we know that we need only move a little in order to find them. In space, on the other hand, outside the Galaxy, our only link with the rest of the universe is through vision, by means of light, and when everything disappears, everything is annihilated.

This void, then, seems to be the prevailing state of the universe in which we live. When we moved away from our solar system, we discovered that day is an unusual phenomenon that occurs only on planets in the vicinity of a star and that the normal condition in space is night. Now we learn that this is true only within the Galaxy or in its immediate vicinity. Outside, even the starry sky gradually disappears, and with it even the type of night familiar to us. Another takes its place: an empty night, composed solely of darkness, a night that seems to annihilate space, divesting it of every dimension—a nightmare sky.

But just as even in the darkest nightmare a glimmer will suddenly appear, enabling one to begin to glimpse something that has shape and substance, and alleviating the sense of oppression, so also in this cosmic nightmare, after having anxiously scrutinized the formless darkness at length, with our backs always turned to the Galaxy, we finally manage to descry a small, faint, nebulous, spindle-shaped speck, which seems to emerge from the nothingness like something at the limit of existence, something that could disappear from one moment to the next, leaving us once more in the agonizing loneliness of the infinite void.

THE ANDROMEDA NEBULA

We also see that nebulous speck from the earth, and it is known as the Andromeda nebula, or M 31 for short. And if, having traversed thousands of light-years from the earth to outside the Galaxy, we still see it as small and faint, we must conclude that it is something very distant and thus, in reality, extremely large and bright. There the darkness must be dispelled anew by light, and new matter must occupy space. Let us therefore direct our journey toward this single glimmer as if it were our last hope.

Little by little as we draw near, we see the speck grow larger and brighter, increasingly rich in details, until, at a distance of 1,200 light-years, it appears to us as we are able to see it from Earth using the largest telescopes. It is a sublime and breath-taking vision that arouses our wonder and admiration. The evanescent light we glimpsed upon leaving the Magellanic Clouds has been resolved into a vast quantity of matter, light, gas, dust, and, once again, hundreds and thousands of stars, differing in brightness and color, single, multiple, or in clusters, constant or variable (Fig. 8.5).

Thus even at great distances extragalactic space is not completely empty. We do not yet know the significance of the agglomeration that we have just reached; we do not know, in short, whether it is an island or a continent. But it is a vast, complex collection of bodies, and we must now discover its real dimensions and nature.

The problem of determining the distance was solved, once again, by the discovery of variable stars of known luminosity. In particular, numerous Cepheids and novae were discovered. By determining the absolute magnitudes of the former from the period-luminosity ratio, and bearing in mind the mean magnitude at maximum of novae observed in the Galaxy, it was possible to obtain at once, from the apparent magnitudes of the Cepheids and the novae, the distance modulus, and from that the same distance in light-years. All this depended on the hypothesis (well founded in fact) that the Cepheids and novae in M 31 behave like those in the Galaxy, especially with respect to luminosity.

It turned out that the Andromeda nebula is 2,300,000 light-years distant from us. This is a distance so great that not only would it make no sense expressed in kilometers but it cannot be placed alongside the other known distances even if it is expressed in light-years. To be able to conceive it, we once more have recourse to a reduced-scale model.

The scale that we adopted to place the solar system within the Galaxy is no longer adequate. Thus let us reduce everything so that 1 cm represents 100,000,000 km. On this scale the sun is a pellet scarcely .14 mm in diameter and the earth a speck invisible to the naked eye, revolving around it at a distance of 1.5 cm. A little over 4 km away we find the nearest star, represented by another pellet, or rather two pellets 0.14 mm in diameter, and the entire Galaxy is contained in a disk 92,570 km in diameter. The two Magellanic Clouds have diameters of 30,850 and 13,270 km, and are located at distances of 154,285 and 185,140 km, respectively.

If we now wish to reach M 31, we shall have to pass through an empty space 2,160,000 km wide. That is an enormous distance, more than five times the actual distance from the moon to the earth. Let us not forget, however, it is a distance measured on a reduced scale, in which the

Fig. 8.5   A series of pictures of the Andromeda nebula. *Left:* Andromeda photo-graphed with a modest astrograph; it is nearly the way we should see it just after leaving the Galaxy, in which case the background would be completely dark, that is, without all the stars belonging to our Galaxy that appear in this photograph. *Center:* Andromeda photographed with the large Schmidt telescope at Mount Palomar; it is seen at its full extent and shows broad spiral arms, gaseous nebulae, and very many clusters. Clearly visible beside it are two minor galaxies, the two satellites that look elliptical. *Right:* A small portion of the Andromeda nebula partially resolved into stars; the photograph was taken by W. Baade with the 200-inch reflector at Mount Palomar. (*L. Baldinelli; Mount Wilson and Palomar Observatories*)

earth-moon distance corresponds to scarcely .004 mm. These, then, are the true proportions of the distances. In other words, on the basis of these results and using these proportions, we can state that of the 2,160,000 km separating us from the Andromeda nebula, man has materially traversed the first .004 mm and, with the probes he has constructed, venturing as far as Jupiter, the first 2 cm.

One immediate consequence of this distance strikes us at once. When we observe M 31, we do not see it as it is now, nor even as it was 2,000 years ago (as with the stars in Orion), but as it appeared 2,300,000 years ago. We are constantly observing so many phenomena there, but all of them occurring at that time, which seems to us so remote. We see pulsating stars, but perhaps by now they have ceased pulsating. We see stars that appear and disappear; perhaps, since the moment when the explosion that we are seeing occurred, they have exploded again many other times and are now quite extinct. Perhaps, during the past two million years, impressive events have occurred that at this point we cannot know about, and if all the stars of M 31 were extinct, or if an immense, inconceivable explosion had torn the entire system asunder, dispersing the stars and the residual matter in space, we should continue to see the usual sight for years and years, until the light arrived to show us all these new events.

But there is another aspect no less interesting. Let us observe M 31 with a telescope or simply with the naked eye (for it is clearly visible without a telescope) and consider that the light that is reaching our eyes at this moment left 2,300,000 years ago, when the appearance of our world was quite different from what it is now; the later Ice Ages had not yet occurred and had not changed various parts of the surface, which was inhabited by animals unlike those in existence today. When that light left, man did not yet exist on the face of the earth.

If, with an effort, we imagine that one of our distant forebears, a hominid of the genus Australopithecus, raised his eyes to the heavens 2,300,000 years ago and noted that faint speck, we will not think that he suspected he was not seeing it as it was at that moment. In fact, the vision of what was happening there in that instant was destined for a species that did not yet exist—we who now see, and perceive as current, events that occurred hundreds of thousands of years before our prehistory.

Sometimes, when we visit the remains of past civilizations or read ancient texts, a tool that we still use or a very modern thought gives us a vivid sense of the relativity of time. Then the Caesars do not seem much more distant than Napoleon, or Phidias than Michelangelo, or Imhotep, creator of the Sakkara pyramid, than Phidias.

But with the sight of M 31 even this sensation is lost. Now it will no longer seem to us as though Caesar and Alexander the Great were on Earth only yesterday, that the construction of the Egyptian pyramids is quite recent, and, moreover, we shall no longer feel that man has existed since time immemorial; rather, it will seem as if we were living in a timeless world where everything past and present is contemporaneous, for there is always some part of space where what happened here long ago appears to be happening now.

A nova suddenly appears in M 31. We have discovered it in a photograph taken the previous night. We continue to observe it every night; we estimate its brightness and trace its light curve, which we then study and discuss. This same nova that we see shining now was discovered over two million years ago by astronomers living on planets in the Andromeda nebula; they traced the same light curve that we are now reconstructing, formed theories, and explained, perhaps, many phenomena that are still not clear to us. These astronomers died (perhaps their species even became extinct), but the same nova continued to appear before the watchful eyes of so many other scholars, at ever greater distances, who recorded its image with as many devices, until it appeared to us, contemporaneous with our own events—with the birth of a child, with a war or a change of government. And in hundreds or thousands of years, when we and our descendants have disappeared, along with all our science (including the light curve that we are now tracing so carefully, for publication in the memoirs of a famous academy, whereby it is transmitted to our contemporaries and to posterity), other beings, more distant from M 31 than we are, will observe the appearance of this nova, will connect it with the current date on their calendar, and will study it scrupulously, as we are doing today. And that ray of light links us all: dissimilar beings, extremely far apart in space, living at different times, who, though performing the same task, are unaware of one another, not just because of the impossibility of communication but also because we operate in different epochs, when others are already dead or do not yet exist.

These reflections on the consequences of the enormous distance of the Andromeda nebula should not make us forget the fact that, now that we are acquainted with it, we can calculate its true dimensions. Having obtained the extreme apparent dimensions by means of delicate measurements on long-exposure photographs, we find that M 31 has a diameter of 165,000 light-years. In the direction perpendicular to the maximum diameter, the length is considerably less, but this is only because of a distortion of perspective, due to the inclination of the equatorial plane to our line of sight. Thus the Andromeda nebula appears to be a fourth astral city, this time similar to our galaxy, even in its dimensions. To make sure, let us try to learn more about it.

Let us start with the variable stars. Up to 1950 some 50 variables had been discovered in M 31, the majority of them pointed out by E. P. Hubble in 1929. In the fifties W. Baade began a systematic search with the tele-scope at Mount Palomar, which is 200 inches in diameter (the largest in the world), examining only four suitably chosen zones at increasing distances from the central zone. The search, completed by others after Baade's death, led to the discovery of 730 variables. Of these, 10 are novae, 439 Cepheids, 65 eclipsing variables (primarily of the β Lyrae type), and 102 irregular variables; the rest belong to other types or have not been classified. Clearly, the total number of variables in M 31 must be much higher, since these variables are only those present in the four zones and, moreover, the brightest of those.

The search for novae has been carried out successfully at various times. Hubble had already listed 85, which appeared in the period 1907–17. The systematic search was later resumed by H. C. Arp, who discovered 30 in the years 1953–55, and by L. Rosino, who found 90 between 1955 and mid-1970. The total number of novae observed in the Andromeda nebula up to the end of 1971 is 224, more than have been observed so far in the Galaxy. According to this research it seems reasonable to assume that on the average thirty appear each year.

Variables represent only a portion of the interesting objects that inhabit M 31. As in our galaxy, here also are found nebulae, associations, and star clusters. Through examination of the best plates taken with the largest telescopes in the world, 688 emission nebulae of the Orion nebula type, 188 O-B stellar associations, and 266 clusters have been found. Open and globular clusters are also found here. Distinguishing between the two types is very difficult, however, because on account of the enormous distance, their apparent diameters are generally greatly reduced, and the details that would enable one to distinguish clusters of one type from those of the other disappear. The distinction has been made, therefore, primarily on the basis of color. The most complete research in this direction is that published in 1965 by M. Vetesnik, who positively identified 257 clusters, of which 167 are, according to him, definitely globular. Their most common absolute magnitude is -8, the same as for the globular clusters of the Galaxy.

To these objects (of which, let us not forget, we see only a fraction) must be added gas, dust, and, particularly, neutral hydrogen, which we can trace by means of radiotelescopes. All this matter, scattered or grouped in clusters or associations, is distributed in a more compact and uniform central zone, from which issue enormous spiral arms, flung toward the outer edge, where they appear to dissolve. According to Baade it is possible to distinguish clearly seven spiral arms. The first two are composed of dust; in the third there also appear supergiants and emission nebulae; the fourth and the fifth contain gaseous nebulae and associations of young stars in great abundance; the outermost two are traced by scattered groups of supergiants. The dust, which by the third arm is already not very abundant, is completely negligible in the four outermost arms.

Like the Galaxy, the Andromeda nebula also rotates rapidly on its axis, at velocities that vary gradually as one proceeds outward. According to the most recent measurements, made by V. C. Rubin and W. K. Ford, the central zone rotates rapidly, reaching a maximum velocity of 225 km/s at a distance of 1,300 light-years from the center. Immediately afterward the velocity begins to diminish, decreasing to a minimum of 53 km/s at 6,500 light-years. As we move further from the center, we find that the velocity increases again, reaching a maximum value of 272 km/s. Farther still toward the peripheral regions, the rotational velocity slowly diminishes, stabilizing just barely above 200 km/s. Thus the stars located in the outermost zones—for example, at 70,000 light-years from the center—make a complete revolution around the center of the system every 660 million years. As in our galaxy, here also gas is observed escaping from the nucleus, at a greater velocity the closer we come to the nucleus itself.

Given the velocity of rotation, it was possible to obtain the total mass of

M 31, which turned out to be 220,000 million times that of the sun. Assuming that half the mass is that of interstellar matter (gas, dust, etc.) whereas the other half is concentrated in the stars, and supposing that the average mass of the stars is that of the sun, we find that there must be more than 100 billion stars in M 31. If each of these were accompanied by some ten planets, as is the case in our solar system, in M 31 alone there would be a million million planets.

This is, in synthesized form, what the observations and their interpretation have taught us so far about the Andromeda nebula, which, from its shape, dimensions, stellar content, and dynamics, is clearly another galaxy similar to ours—indeed even bigger and fuller.

The similarity extends even further, for, just as the Galaxy has two minor satellites (the Magellanic Clouds), M 31 also has two much smaller companions, elliptical in appearance, which are designated in the celestial catalogs by the symbols NGC 221 (or M 32) and NGC 205.

THE WORLD OF GALAXIES

Having discovered the Andromeda nebula, we should stop for a moment to reflect and to look around us. Since leaving the Galaxy, we have discovered five other systems similar to it, even though all except one are smaller. Since these six galaxies exist and are separated by vast empty spaces, we might suspect that if we were to traverse other, equally vast spaces, we would encounter other galaxies. The systematic observation of the sky by means of the largest telescopes (carried out especially in the last half-century) has fully confirmed this supposition, showing us, at increasingly greater distances, tens, hundreds, thousands of galaxies. Thus space beyond our galaxy is not empty: it is populated by countless galaxies, often separated from one another by distances that are enormous compared to distances between their stars and inconceivable in our galaxy, but not excessive compared to the dimensions of the galaxies.

A few decades ago, when we were just beginning to acquire some idea of the dimensions and distances of the galaxies, many astronomers referred to them as "island universes." Although somewhat rhetorical, this term does express the reality, since every galaxy appears as an autonomous entity that has everything: dark and shapeless matter, stars being born, dead stars, and stars at the peak of life, many accompanied by planets; dust and gas, cold bodies and blazing stars, dark specks and faint comets, very abundant aggregations of stars, and often, between them, immense voids. In sum, a galaxy is comparable to an immense city, with its streets, houses, parks, and cemeteries—a city of stars instead of houses.

Though retaining these common characteristics, galaxies, like cities, can differ greatly from one another in appearance and dimensions. This variety, which we have barely glimpsed among the first six, has become increasingly evident with the rise of the number discovered. Today we know of larger and smaller galaxies, elliptical, spiral, or irregular in appearance, with every possible variety in between; some have nuclei that are stellar in appearance, or others are reduced to a nucleus alone. Other very strange-looking galaxies have been discovered: galaxies in pairs and even in groups. In sum, we have found that there exists, on a scale much vaster than the stellar scale, a new world to be explored, the world of galaxies, and this, finally, is perhaps truly *the* world.

To explore it requires the greatest mental effort and a boundless flight of fancy, but we shall penetrate it—at least as far as we are able. Science can no longer guide us to a boundary, as it did for the solar system or the Galaxy. We shall travel until we reach its limits—our own limits—and when we are obliged to come to a halt, it will be because we have been stopped, not by certain problems, great or small, that we can leave behind us (as has already happened many times) but by the problem that, in a way, sums them all up: the mystery of the limits of the cosmos, its origin, and its evolution. Because now—and only now—we are about to tackle the problem of the universe.

Let us start with more common galaxies, similar to those we already know, and which we shall be tempted to call "normal." Normal galaxies are of three types: spiral, elliptical, and irregular. We have already encountered an example of each of these types. In fact our galaxy and M 31 are of the first type, the two satellites of M 31 belong to the second type, and the Magellanic Clouds to the third. The spiral galaxies fall into two large subgroups: normal spirals (S) and barred spirals (SB). In the first subgroup the spiral arms seem to issue from the central zone just like a logarithmic spiral; in the second they seem instead to start from the two ends of a kind of bar that crosses the nucleus. Each of these two subgroups is divided into three classes, indicated by the letters *a, b,* and *c,* according to whether the spirals appear more or less developed.

The elliptical galaxies (E) show no spiral arms or other special configurations and may be completely spherical or more or less flattened. The extent of the flattening is indicated by a number from 1 to 7. Some galaxies that are more flattened, of type E7, but show no traces of spiral arms are indicated by the symbol S0; naturally, there is also a subgroup SB0. The elliptical galaxies have one very strange characteristic: they show no trace of dust, that is, they are composed exclusively of stars.

Finally, the irregular galaxies show no particular structure and are generally smaller than the others.

The principal types of galaxies are illustrated by twelve examples chosen from the numerous photographs taken by Hubble, creator of the classification system just described. Some of the galaxies are seen full face, others in profile (Figs. 8.6, 8.7), depending on how the equatorial plane of each is inclined with respect to our line of sight.

Of all the galaxies known and classified, the spirals (Fig. 8.8) are the most common (78%); of these, only 15% belong to the group of barred spirals (Fig. 8.8, *bottom*). Ellipticals account for 18%, irregulars 4%.

These percentages, however, are not to be taken as definitive, for obviously these statistics favor the brightest galaxies, those we can most easily discern at great distances and which we come to know in considerably greater numbers.

The selective effect of our opportunities for observation may be due not only to the intrinsic total brightness of the galaxies but also to other causes that prevent us from recognizing a galaxy as such. Atypical galaxies do in fact exist; they are difficult to observe, but they may greatly outnumber the common galaxies.

DWARF GALAXIES AND COMPACT GALAXIES

Let us begin with the so-called dwarf galaxies. The first of these strange stellar systems was discovered by H. Shapley in 1937. Examining a photograph of the southern constellation Sculptor, he was struck by a vast aggregation of scarcely perceptible points. The strange configuration resembled a

Fig. 8.6   The galaxy M 101, one of the most extensive among the galaxies that we can see face on with its well-developed system of spiral arms. Here too, as in Andromeda, we can see bright nebulae, dark nebulae, and star groups. (*Mount Wilson and Palomar Observatories*)

Fig. 8.7   The galaxy NGC 891, seen in profile, resembles the wide-field photographs of the Milky Way, suggesting that the Galaxy, seen from outside and in its equatorial plane, would appear more or less the same. (*Mount Wilson and Palomar Observatories*)

fingerprint on the plate and could have consisted, in fact, of a collection of spurious points, due solely to the emulsion. On the other hand, the plate was particularly sensitive, and the object was not visible in other plates taken with the same instrument. After examining hundreds of plates of the region, Shapley finally succeeded in finding one (taken thirty years earlier, with an extremely long exposure) in which the object was visible.

Once the existence of such an object had been confirmed, there remained the task of deciphering its nature. The group of points could have been a nearby star cluster made up of very faint stars, or else a group of extremely distant normal stars, or a cluster of galaxies so distant that its components, photographed with a small instrument, appeared as points, like stars.

A plate taken with the 60-inch telescope of Harvard Observatory soon excluded the last possibility: the object turned out to be an aggregation of stars, and, on an apparent surface of about two square degrees, at least ten thousand could be counted. Subsequently, by comparing various photographs taken with the 100-inch telescope of Mount Wilson (then the largest in the world), more than forty Cepheid variable stars of the RR Lyrae type were found in the region. Since the absolute magnitude of stars of this type is well known, it was easy to obtain their distance (that is, the distance of the system to which they belong), which turned out to be 280,000 light-years. The object discovered by Shapley was thus an aggregation of stars outside our galaxy, much larger than a star cluster but much smaller than a normal galaxy.

While the object in the constellation Sculptor was being studied, hundreds of plates of other regions of the sky were examined at Harvard Observatory in an attempt to find other similar objects. The search led to the discovery of another such object in the constellation Fornax. Associated with this object, which showed no RR Lyrae variables, were a number of globular clusters (of which six are so far known). From the luminosity of their brightest stars a distance of over 600,000 light-years was obtained. Thus this object is also extragalactic and much more distant than the first. Further searches for stellar systems of this type had negative results until the great Mount Palomar photographic atlas was produced (between 1949 and 1956). In the course of a study of it four other such objects were discovered: two in the constellation Leo, one in Ursa Minor, and one in Draco.

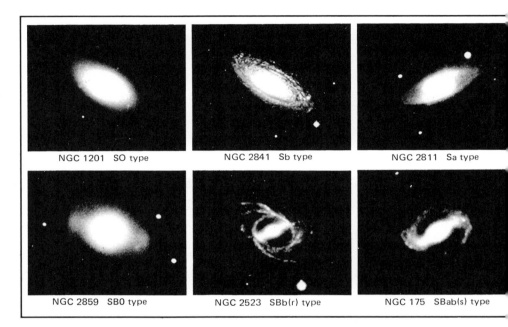

NGC 1201 SO type   NGC 2841 Sb type   NGC 2811 Sa type

NGC 2859 SB0 type   NGC 2523 SBb(r) type   NGC 175 SBab(s) type

Fig. 8.8 Two sequences of spiral galaxies of various types: *top,* spiral galaxies; *bottom,* barred spiral galaxies. Both are classified according to the nomenclature formulated by E. P. Hubble. A spiral galaxy (S) show arms which seem to issue directly from the central zone, whereas a barred spiral (SB) has arms that seem to issue from the ends of a "bar" that crosses its center. (*Mount Wilson and Palomar Observatories*)

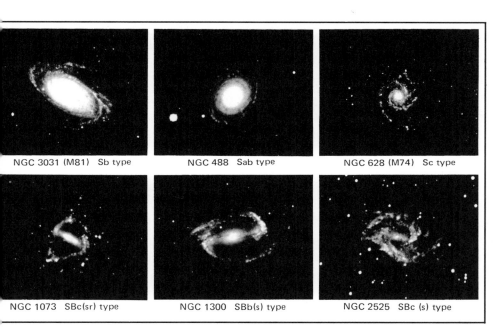

NGC 3031 (M81)   Sb type   NGC 488   Sab type   NGC 628 (M74)   Sc type

NGC 1073   SBc(sr) type   NGC 1300   SBb(s) type   NGC 2525   SBc (s) type

Fig. 8.9   A dwarf galaxy in the Sextant constellation. Note the faint surface brightness and the very disconnected appearance of the stars forming it. (*Mount Wilson and Palomar Observatories*)

The study of these six systems showed them to be stellar aggregations from ten to a hundred thousand times fainter than ordinary galaxies, with diameters between 2,500 and 7,000 light-years. These diameters are enormous compared with those of star clusters but small with respect to those of galaxies. Thus the aggregations could not be considered gigantic clusters or small galaxies. For various reasons (including the fact that the aggregations themselves may contain clusters), the second opinion has prevailed, and they are called dwarf galaxies (Fig. 8.9).

All the dwarf galaxies recorded so far are contained within a radius of 1,300,000 light-years from us. Recalling that within the same radius there are only three "normal" galaxies, so to speak (ours and the two Magellanic Clouds), we immediately conclude that if the same distribution holds in the rest of space, there must be at least twice as many dwarf galaxies as there are normal ones. And, indeed, it is very reasonable to believe that some also exist at distances greater than 1,300,000 light-years and elude us only because it is difficult—or impossible—to observe them.

Moreover, after the discovery of these six dwarf galaxies, spheroidal in appearance, S. van den Bergh found—again, in the Mount Palomar atlas—many others (his 1959 catalog includes 222), brighter and more extensive, but still smaller than the Magellanic Clouds. These galaxies, which vary in appearance (irregular, spiral, etc.), belong to the group of dwarfs, undoubtedly numerous, but difficult to discover and observe.

But it is not only their faintness that can conceal these objects from us. In recent years galaxies of another type that generally escape observation have been discovered: the "compact" galaxies (Fig. 8.10). These, unlike the dwarfs, have considerable surface brightness, but their diameters are so small (always, it is understood, on a galactic scale) that unless they are sufficiently close to us, they appear almost pointlike and are indistinguishable from stars.

One of the most typical was observed at Mount Palomar Observatory by the astronomer H. C. Arp. This object, which appears on photographs taken with the world's largest telescope as a star of magnitude 18 (that is, over a hundred thousand times fainter than the faintest star visible to the naked eye), turned out to be over 40 million light-years away and to have a diameter of scarcely 230 light-years. Thus this galaxy has the same dimensions as the largest globular clusters but is a hundred times brighter. Yet its diameter is hardly one-hundredth of the diameter of the largest dwarf galaxies, and its luminosity is of the same order as that of the brightest.

This strange galaxy is perhaps at the lower limit of the class of compact galaxies, which have been sought and found in recent years primarily by F. Zwicky. Using the Schmidt telescope at Mount Palomar, he has discovered some hundreds of compact galaxies; those discovered earliest have already been studied in detail with the 200-inch reflector. On the basis of this latest research Zwicky has already reached some interesting conclusions, which, being based solely on a few cases, are naturally susceptible to modifications.

The compact galaxies observed thus far are between 30 million and 1 billion light-years distant. Their diameters are between 300 and 7,000 light-years, that is, equal to, or even less than, those of the dwarf galaxies. Their luminosities, however, are exceptionally high, surpassing not only those of dwarf galaxies, but also, in certain cases, those of normal galaxies, whose dimensions are much greater. Thus their surface brightness must be extremely high. Their spectra often exhibit emission lines (sometimes only these) and extremely broad lines, from which Zwicky deduces high velocities within the galaxies themselves and high values for their masses.

This is perhaps the most surprising conclusion. The mass-luminosity ratio in these galaxies is in fact from a hundred to a thousand times greater than that found for the sun. Since, in general, the average mass of stars corresponds roughly to that of the sun, it follows that these galaxies are from a hundred to a thousand times heavier than they would be if they were composed, as is generally the case for other galaxies, only of stars of solar mass. We must conclude, therefore, that compact galaxies contain a great quantity of dark matter. This matter escapes direct observation, but its existence can be deduced by indirect methods, an example of which has just been given.

The study of these strange galaxies presents considerable practical difficulties; moreover, it is only in the early stages. The greatest difficulty, however, lies in the actual discovery of both compact and dwarf galaxies. According to a diagram prepared by Arp (Fig. 8.11), if we examine the apparent diameters and the absolute magnitudes (that is, the total intrinsic brightness), we see that the observable galaxies lie within a relatively narrow oblique band. The lower part of this strip contains the star clusters; the

Fig. 8.10 A compact galaxy discovered by F. Zwicky. (*Mount Wilson and Palomar Observatories*)

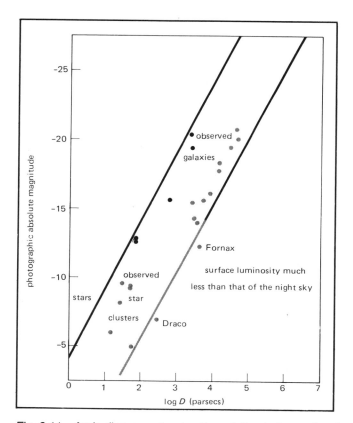

Fig. 8.11 Arp's diagram, showing the relation between the diameter and the absolute magnitude. The area between the straight lines is that in which galaxies can be discovered by photographic means.

upper part the normal galaxies. Between the two groups we see a vast gap in which no object had yet been found: the compact galaxy described by Arp is located in precisely this zone and thus seems to constitute a link between star clusters and galaxies.

It is important to note that any object outside this belt normally escapes our observations: those on the left because, being bright but compact, they are taken for stars; those on the right because, being too large in proportion to their brightness, they are lost in the sky's background. Nevertheless, in rare cases, as we have seen, galaxies may be discovered even in these regions: for example, when compact galaxies (that is, those stellar in appearance) attract our attention by emitting radio waves, or when dwarf galaxies are close enough for us to observe individual stars. The fact that, in particularly favorable cases, galaxies are found even in the empty zones strongly suggests the conclusion that the sole cause for the absence of galaxies in the part of the diagram outside the strip is the insufficiency of our means of observation.

Thus not only has the discovery of compact and dwarf galaxies introduced us to two new types of extremely interesting galaxies, but it also shows us that the intergalactic spaces are less empty than they seemed to be right after we left the Magellanic Clouds. If our vision had been sharper, we should have descried not only the remote Andromeda galaxy but also other galaxies, such as the closer dwarf galaxies or the brighter compact galaxies.

With this survey our voyage in the world of galaxies has run its course. We started out by acquainting ourselves with the more common galaxies, which we may henceforth term classical; now we have discovered new types of galaxies which have been the subject of research for only the last few years. Soon we shall be encountering galaxies and objects that are stranger still, and finally we shall hurl ourselves toward the bounds of the accessible universe, in an attempt to obtain the broadest possible overview of the whole. But before we proceed, we would like to direct our glance once more toward that part of space from which we set out.

Looking out from the Andromeda nebula, we can still discern our galaxy at the limit of visibility with the naked eye. It is a speck slightly smaller and fainter than M 31 as seen from the earth. There in that speck, which is all we can see of the light of billions of stars, is our sun, and near it the earth.

But now we have no hope of finding it. From the distance of M 31 the sun would appear as a star of magnitude 29, that is, a hundred times fainter than the stars we would barely be able to discern in photographs of our galaxy taken from M 31 with a telescope as large as the one at Mount Palomar, the largest in the world. And even if we were able to use a telescope powerful enough to see such faint stars, we should run up against another great difficulty. Indeed, with it we should see not only the brighter stars but also all the other stars at least as bright as the sun, which are very great in number: there are billions of them. Trying to find the sun in their midst would be like trying to find a particular grain of sand on a beach 80 meters wide and 1 kilometer long, down to a depth of 10 centimeters.

It is useless! The earth has long since disappeared, and we must resign ourselves to the disappearance of the sun as well. Soon, as we move farther and farther away, even the Galaxy will vanish, swallowed up in the distance. But how can the sun and the stars of the Galaxy be of any further interest to us now that we are beginning to count more galaxies than there are stars visible to the naked eye from Earth, and now that we are venturing forth in search of the bounds of the universe?

The world of the galaxies—here, truly, is infinity! One who has a yearning for the sky can satisfy it by leaving the earth and going up into the atmosphere; one who aspires to space can find satisfaction in an earth-moon trip or an interplanetary voyage; but one who longs for the infinite can experience its rapture only in the world of the galaxies—there where the universe eludes even our imagination. To set forth among the galaxies, after traversing the solar system and even the entire Galaxy, is rather like flying around the world as compared to stepping outside for a breath of fresh air.

And that is what we are now doing: traveling through these spaces so distant from the spot where our body remains that we have even lost track of the Galaxy. We began by viewing the cosmos; now we are penetrating it, almost as if we were entering a new dimension.

If some of our fellow travelers, confronting extragalactic space for the first time, have been deeply affected by the world of the galaxies and are experiencing vertigo now that they are face to face with the immensity of the cosmos, they would be well advised to stop at this point. Under such conditions, in fact, it would be very difficult for them to grasp fully the new wonders that we are about to discover, and to rejoice at the vision that awaits us, just as someone who has succumbed to mountain sickness during the climb is unable to enjoy the view from the summit. But if, however, not content with what we have discovered so far, they wish to continue, they would do well at least to pause for a while in order to acclimatize themselves, for the final stage of the journey we are about to undertake stands to offer us spectacles that are even more amazing.

Never has the human eye been able to discern anything grander than what we are about to perceive. Never has the imagination dared to conceive objects and phenomena such as we are about to discover. And it is all around us, all the time; if we had sharper eyesight, we could perceive it every night overhead.

Indeed, since from now on it will be practically impossible to continue to follow an itinerary, we shall stop at some point in space that we have already visited: on the Andromeda galaxy, in some isolated part of space, or even on our own planet. It does not matter, for the spaces that we are about to probe are so vast that it makes no difference whether we observe them here or from M 31. From this base we shall extend our exploration to the outermost limits of our instruments and our minds.

We shall continue to concern ourselves only with what is most significant; we shall examine only samples of the objects and phenomena we encounter, leaving to our imaginations the task of multiplying them as the occasion arises, so as to provide an overall view.

EXCEPTIONAL EVENTS

A truly exceptional specimen is the one we shall encounter right away: M 87 (Fig. 9.1). Here we have an elliptical galaxy of the EO type (that is, perfectly spherical), which at first glance would appear quite normal: round, not that large in diameter, and of an average brightness. Its only peculiarity is a thin, rectilinear jet of matter that issues from its center in a north-west direction (Fig. 9.2). For those who want the exact figures, let us note that the appar-

Fig. 9.1   Wide-field photograph of the group of galaxies in the Virgo constellation, to which M 87 belongs. *From the center toward the bottom,* the galaxies NGC 4438, M 86, and M 84, the last of a long chain (note the unusual shape of the first); *above, slightly to the left,* the giant elliptical galaxy M 87. The photograph was taken with the Schmidt telescope at Mount Palomar, by means of a special technique. (*F. Bertola*)

Fig. 9.2   The central zone of the M 87 galaxy, photographed with a short exposure so as to show the jet of matter issuing from the center. (*Mount Wilson and Palomar Observatories*)

ent diameter of the galaxy is 3', and the jet, directed at a position angle of 290°, is 20" long and 2" wide. Of course, that is how it appears from the earth. In fact, the real dimensions depend on the distance, and the apparent length of the jet would correspond to its real length only if it were oriented perpendicular to our line of sight. If, however, it is somewhat at an angle with respect to us, then what we see is only its projection, and it could actually be much longer.

Increasingly detailed observations, carried out especially in recent years and still in progress, have shown that the rectilinear jet is not the only strange characteristic of this galaxy, and perhaps not even the most unusual.

By late 1949 it was discovered that M 87 emits radio waves. The singularity of this lay not so much in the fact itself (which became clear later on, after many galaxies that emit radio waves had been discovered), but in the nature of the emission. Subsequent research showed that the radio emission takes place essentially in three components: an extended one, elliptical in appearance, of apparent dimensions 6' by 10'; an intense one, along a thin band about 50" long; and a central one, almost pointlike, which, according to interferometric measurements carried out in Australia and California, has a diameter of scarcely 3 light-months.

In 1965 instruments installed on an Aerobee missile, launched to study celestial X-ray sources, demonstrated that M 87 is a powerful emitter of X rays, which it sends into space with an intensity from ten to a hundred times greater than that of its optical and radio emission. At that time no other extragalactic X-ray sources were known; today some tens of sources are known, of which only nine, among them M 87, have been positively identified.

However, the surprises were not over in the optical field. As we have already seen, galaxies are often accompanied by tens or hundreds of globular clusters, arranged around them like a halo. This occurs in our galaxy and in M 31, and it is not strange that it should also happen in M 87. Long-exposure photographs have indeed revealed this ring of clusters, but in enormous numbers: many hundreds (Fig. 9.3). Over a thousand have been counted, and, considering the fact that the fainter ones escape us, we can calculate that, if the bigger ones are of the same magnitude and intrinsic brightness as the largest ones in our galaxy or in the Andromeda galaxy,

Fig. 9.3   The M 87 galaxy photographed with a long exposure. The luminous points visible in the outermost zones around it are not stars but globular clusters, each of which is made up of hundreds of thousands of stars. (*Mount Wilson and Palomar Observatories*)

the total number must be about 2,100. To get an idea of how enormous such a number is, we need only recall that in the whole of our galaxy we have found little more than a hundred clusters, and in M 31 less than two hundred.

From the apparent brightness of the globular clusters (still supposing that the most luminous have the same luminosity as the most brilliant clusters in the Galaxy or M 31), the distance of M 87 was obtained; it had already been determined through other, less precise methods. According to the measurements of A. Sandage (1968), which are in close agreement with subsequent measurements, M 87 is located at a distance of 48 million light-years.

Now that we know the distance, we can determine this galaxy's true dimensions. Here again the most recent observations have brought us a great surprise. Using the largest telescope in the world and special photographic techniques, H. C. Arp and F. Bertola discovered that M 87 is actually much larger than it has appeared so far, for the fainter outer zones always escaped us. Its diameter turns out to be almost one degree, corresponding to a real diameter of 848,000 light-years. Thus we are dealing with a gigantic galaxy, very much larger than our own or the Andromeda galaxy, which previously were among the largest known.

But a result that is even more astonishing has been obtained for the mass. The most recent determinations, undertaken particularly after the discovery of the large diameter, have led to estimates on the order of millions of millions of times the mass of the sun. According to the most recent research, published in April 1970, M 87 weighs 27 million million times more than the sun. Of course, this does not mean that it is necessarily composed of 27 million million stars like the sun, for it may be that much of the matter is still amorphous. However, it is equally impressive to reflect that, with the matter in M 87, it would be possible to construct over two hundred galaxies as large as our own galaxy, or a hundred like the Andromeda galaxy, or fully forty-five hundred smaller galaxies like the Small Magellanic Cloud.

But let us come to the most sensational discoveries, which primarily concern the central jet, noted at the start. This strange configuration, pointed out as early as 1918, has been studied more thoroughly only in the last twenty years, and especially in the past few years. It has an average diameter of 470 light-years and a length of at least 5,000. It is not, however, certain that it is a perfectly continuous and homogeneous cylinder, since, even when observed from the enormous distance at which we are located, it shows interruptions and knots. Both the jet and the knots are extremely luminous. G. de Vaucouleurs has shown that the latter appear almost pointlike and must have diameters no greater than 60 light-years; on the other hand, the brightest of these turns out to be sixty-five times more intense than the most brilliant globular cluster associated with M 87.

One of the most important discoveries concerning the jet was made in 1956 by Baade, who found that the light emitted is polarized. Thus the jet of M 87, like the Crab nebula, the famous remnant of an exploded supernova, also emits synchrotron radiation, as the Russian astronomer Shklovskii had already predicted a year before Baade observed it. Thus the energy dispersed in space in the form of light and radio waves is emitted by relativistic electrons, that is, by electrons moving at speeds approaching that of light, accelerated by a magnetic field; and in this case the energy in question is enormous. Taking the estimate made by G. R. Burbidge (based on the well-known theory of synchrotron radiation, tested in terrestrial laboratories), one finds that in less than a year the jet of M 87 emits, as light, a quantity of energy equal to the whole of that produced and emitted by a supernova during its entire existence.

This means that not only is an enormous amount of energy produced there, but that, for this very reason, we must be dealing with a temporary phenomenon that cannot last long. In fact, an attempt was made, still based on the theory of synchrotron radiation, to determine the average lifetime of the electrons responsible for the optical emission, and it was found to be on the order of a thousand years. Of course, over the life of the jet there may have been various generations of optical electrons, but that does not change the fact that the jet itself cannot have lasted much longer.

In order to try to learn something more, let us move along the jet toward the interior of the galaxy and try to see what is now happening in the nucleus. Spectroscopic observations have shown that it is very rich in gas, which is mostly ionized oxygen. This gas, however, is not at rest or in turbulent motion but is expanding very rapidly, moving away from the center at a velocity of 900 kilometers per second. If we assume that the jet is also formed of material expelled from the center at such a velocity, we can easily calculate that it must have emerged about 15,000 years ago. Of course, the time might be greater if the jet were longer than it appears (let us not forget that we see it in projection), or less if the gas of the jet were expelled at a velocity higher than that now observed in the nucleus. Given that both of these considerations are almost certainly valid, if one of the two effects does not clearly prevail over the other, they tend to counterbalance each other, and the age of the jet can be roughly estimated at 15,000 years. At any rate, independently of more precise measurements, the following now seems certain: that the jet has existed for only a few thousand years (a paltry amount of time if we consider its dimensions and recall that the evolution of a galaxy is measured in billions of years) and that it was formed by some kind of explosion that took place in the nucleus.

We might ask how this explosion took place, and why only once and in only one direction. Although we have no clear ideas about the causes of such a phenomenon, we do know that, according to recent observations, there exist other, fainter jets, proceeding from the center in other directions; indeed, one of these is exactly opposite the principal one. Thus it seems there was more than one explosion. But at this point, this kind of phenomenon can best be observed in another galaxy that provides even more suggestive evidence.

THE M 82 GALAXY

This is M 82, a galaxy seen in profile in the constellation Ursa Major, not far from a beautiful spiral galaxy, M 81. Its mass is hardly one-thousandth that of M 87; it is also much less extensive. Nevertheless, since it is much closer to us (10,500,000 light-years) and of the irregular type, it appears fairly large and, above all, very rich in details. Here and there are seen various bright condensations and many dark lines that cross it in various directions, revealing the presence of a considerable amount of interstellar dust. Infrared photographs have revealed the presence of a compact nucleus, close to the center. This nucleus never appeared in photographs taken at visual wave-lengths, since it was masked by the enormous quantity of dust, which is, however, transparent to infrared radiation. But the greatest discovery, effected by using special photographic techniques, was that made in 1963 by R. C. Lynds and A. Sandage. Here is how things went and what they discovered.

When M 82 was observed visually or photographed without special devices, it always looked like a normal galaxy seen in profile; slightly spindle-shaped, with the details that have been described (and as it appears in Fig. 9.4, *top*). This image is due to the light emitted by the galaxy's gases and especially by its stars, which of course we cannot distinguish individually, because of the enormous distance. Lynds and Sandage, however, photo-graphed M 82 on a plate particularly sensitive to the red light of the Hα line of hydrogen and with a filter that transmitted the light emitted by that line almost exclusively (Fig. 9.4, *bottom*). The result was an image completely different from the usual one. From the central zone emerge two enormous projections, perpendicular to the equatorial plane, which are due to the light emitted, not by the stars but by gas, that is, by hydrogen. These hydrogen projections consist of two enormous systems of filaments (one on each side, with respect to the equatorial plane) with a maximum length of 10,000 light-years. Spectroscopic observation of these filaments has shown that the gas of which they are composed leaves the nucleus at a velocity of 1,000 kilometers per second.

Thus we find in this galaxy a phenomenon analogous to that observed in the nucleus and jet of M 87, except that here it does not occur in only one direction but is an immense explosion involving a whole galaxy. Enormous quantities of matter in the gaseous state are moving away from the center at the fantastic velocity of 1,000 kilometers per second. In order to form an idea of the violence of such an event, we need only recall that the terrible explosive wave caused by an atomic bomb collides with matter at a velocity of 75 kilometers per second, and that happens only in the first few instants and in the area immediately adjacent to the explosion: the shock wave that reaches the objective, sweeping away houses and buildings like twigs in a wind, is already much weakened. Therefore, the interior of this galaxy is exploding with a violence at least twenty times greater than that of an atomic bomb. And we are only dealing with the violence of the explosion, not the energy liberated, which obviously is incomparably greater, given the enormous quantity of matter that is exploding. It has been found that the energy released is 10^{55} ergs, that is, equal to the energy of a million super-novae. And let us not forget that the number of atomic bombs required to

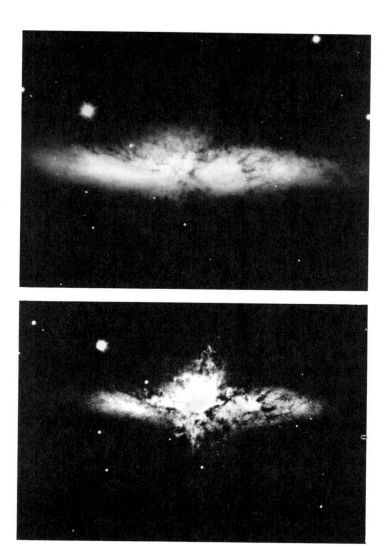

Fig. 9.4   The M 82 galaxy in the constellation Ursa Major, which is 10.5 million light-years distant from us and whose mass is hardly a thousandth of that of M 87; *above,* photographed with a normal blue plate, M 82 does not differ much from any · irregular galaxy seen in profile; *below,* the same galaxy, photographed in red light (so as to record the emission corresponding to the Hα line of hydrogen), shows two enormous cones of gas, expelled from the center at very high velocity in a colossal explosion that took place at most twelve million years ago. (*Mount Wilson and Palomar Observatories*)

match the energy of a supernova is the number 1 followed by twenty-eight zeros! In other words, it is as though 10^{34} atomic bombs were exploded in the center of M 82.

From the velocity of expansion of the filaments (reversing the process), it is found that the explosion must have begun a maximum of 1.5 million years ago. I say "a maximum" because the calculation was made on the basis of the current velocity, whereas the initial velocity was almost certainly greater, because the expanding gas must have been slowed down by colliding with the gas already present in the galaxy. Therefore, a maximum of 1.5 million years ago, an explosion occurred in the galaxy M 82 that is still going on.

But wait: this is what *we* see. In reality, since M 82 is 10,500,000 light-years away, we can only say that 12 million years ago an explosion began that we see still spreading outward after 1,500,000 years. Since then over 10 million years have passed. What may have occurred on M 82 in the meantime? How will this galaxy have been transformed? What would it look like now if we could observe it close up? These are questions that we shall never be able to answer, and our descendants, who would be able to if they ever exist, are much farther from us in time than we are from Stone Age man.

But it is useless to torment ourselves, regretting what we can never know. There are still so many mysteries to unravel, so many interesting discoveries to make concerning what we can see right now.

And it is just on these grounds and for these reasons that we promise to show our fellow travelers, presently, something even more amazing than the spectacle the M 82 galaxy may present to the inhabitants of Earth in 10,500,000 years.

THE SEYFERT GALAXIES
Meanwhile, we should continue our voyage among the galaxies in whose interiors are unleashed events of an extraordinary violence, never before witnessed.

Apart from a few other cases roughly similar to the two described, we know of a certain group of galaxies (scattered in various parts of space but with similar characteristics) in whose nuclei tremendous explosions are actually being observed, which perhaps resemble the explosion in M 82, as it would have appeared in its early stages. These are the so-called "Seyfert

Galaxies," named after the American astronomer who discovered the first few in 1943. Seyfert never realized the full import of his discovery, for seventeen years later he lost his life in an automobile accident. How great and yet how feeble is man, who manages to know so much and can be destroyed so easily!

Today we know of a dozen Seyfert galaxies (according to the original classification). When observed directly, they appear quite similar to normal spiral galaxies, except for one strange detail: the center is occupied by a very brilliant, pointlike nucleus, similar to a star. This was the characteristic by which Seyfert distinguished them. But a more thorough study, carried out in recent years with various observational techniques, has revealed other, even more important features.

Given the immense distance that separates us from these objects, the nucleus, despite its stellar appearance, cannot actually have the dimensions of a star; that is, it is only small relative to the rest of the galaxy to which it belongs. Spectroscopic observations have shown the presence of intense lines of hydrogen and doubly ionized oxygen, in emission and in highly excited states. The considerable broadening of these lines indicates that the gases present in the nucleus are expanding at a velocity of about 5,000 kilometers per second. Moreover, M. F. Walker discovered that in one of these galaxies the nucleus consists not of a single mass of gas, but of at least four clouds, expanding at different rates.

A second important characteristic concerns the color of the light emitted. The light of normal galaxies is predominantly that of stars of the solar type (of which they are mainly composed)—that is, yellow. This also holds for a Seyfert galaxy, if we take the light it sends us in its entirety. If, however, we limit the light that reaches our instrument by means of a diaphragm centered on the nucleus, and if we steadily contract the diaphragm, the color shifts progressively toward the blue, the violet, and the ultraviolet. Obviously, this means that, whereas the rest of the galaxy is composed of stars, most of them more or less like the sun, its nucleus emits mainly blue-violet light. The Seyfert galaxies are also radio emitters, generally very weak and often with a variable intensity.

But the most baffling news concerning these objects was that announced in 1967 at the Prague International Congress: the pointlike nucleus of some Seyfert galaxies appeared variable. This news caused a great sensation,

primarily because, given its great distance from us, even a pointlike nucleus like those described must have enormous dimensions, and the greater
the dimensions, the harder it is to explain the variability, as the phenomenon involves an ever greater quantity of matter. This mystery has not yet
been solved. Perhaps it is linked to the phenomena of expansion observed
spectroscopically. At any rate, however mysterious remain many of the
phenomena observed in these galaxies, especially the variability of the
nucleus, one thing is certain: even in this instance we are observing true
explosions; and further, given that the velocities of expansion can reach
5,000 kilometers per second, they are up to five times more violent than the
explosion observed in M 82. Perhaps the Seyfert galaxies show the actual
initial phase of the explosion, more circumscribed but more violent, and M
82 its subsequent evolution.

From the dimensions of the nucleus (measured directly) and the velocity
of expansion of the gas (observed spectroscopically), it is possible to determine the length of time needed for the gas issuing from the center of a
Seyfert galaxy to form a nucleus of the diameter currently observed. This
turns out to be an interval of about 20,000 years. Thus we are clearly
dealing with explosions much more recent than those of M 82.

But calculations can also show us the time that must elapse before all the
energy currently present in the form of kinetic energy (that is, as the energy
of matter in motion) is radiated into space in the form of emission lines,
such as the observed lines of hydrogen and oxygen. It turns out that about
3 million years are required. This time could be shorter if, for example, the
energy were also dissipated in other forms (and we have seen that this can
happen by means of radio emission), but it is significant that the order of
magnitude of the duration of the phenomenon is a million years, that is, the
length of time that the explosion in M 82 has already lasted. Even the
energy associated with the exploding mass seems to be of the same order
of magnitude as the mass liberated in the explosion of M 82: about 10^{55}
ergs.

From all these observations there clearly emerges an unusual view of the
world of galaxies, quite different from the one we have hitherto had. It is no
longer solely the realm of vast distances, of tremendous agglomerations of
matter arranged symmetrically in the absolute peace of the cosmos; it can
also be the site of terrible catastrophes, where the violence exceeds the

limits of our imagination, as was already the case for the masses, the distances, and the number of stars.

For the moment we do not know much more than what we have seen. But that is all we need in order to outline a preliminary sketch of what must happen. In a few galaxies, which may be elliptical (like M 87), irregular (like M 82), or spiral (like the Seyfert galaxies), violent events can occur, in the course of which as much as 10^{56} ergs of energy can be emitted. The effects of these events are manifest in various ways and persist up to a million years or more after the beginning of the explosion. During the explosion great quantities of relativistic particles are emitted (at extremely high velocities); their presence is revealed by the emission of radio waves and polarized light. Finally, matter is ejected into space, at enormous velocities.

Velocities greater than or equal to 1,000 kilometers per second had already been observed on a few rare occasions, as in the remnants of supernovae or in certain types of solar activity, but these cases involved relatively very small masses. In the Seyfert galaxies, where the event is presumably observable from the start, the quantity of matter moving at velocities between 1,000 and 3,000 kilometers per second has a total mass between a million and ten million times that of the sun. The escape velocity from the center of a normal spiral galaxy (that is, the velocity that must be imparted to a body for it to escape into space and not fall back to the galaxy) is about 500 kilometers per second, whereas the velocity of escape from a more massive elliptical galaxy is about three times greater. Thus the matter expelled could escape from the galaxy, as long as it was not held back by colliding with other matter already present. The escape depends, therefore, solely on the amount of the gas already present and its distribution. Thus in a flattened galaxy (such as a spiral or irregular galaxy), if the explosion expels a mass ten million times that of the sun from the nucleus, at an initial velocity of 3,000 kilometers per second, the matter hurled in the direction of the disk will, after having traveled 30,000 light-years, be slowed down to a velocity of barely 50 kilometers per second, whereas the matter launched in the perpendicular direction will encounter only slight resistance, move rapidly beyond the galaxy, and continue to spread at a high velocity. Of course, the velocity will be lower, the more the direction of the expansion deviates from the perpendicular to the equatorial plane of the galaxy, since the thickness to be penetrated, and hence the resistance, increases. In practice, in a spiral or irregular galaxy the matter expelled will move out along two cones. That is precisely what is shown by the two projections of gas observed in M 82. However, measurements of an explosion observed in another galaxy with spherical symmetry (NGC 1275) have shown the same velocity, about 3,000 kilometers per second, in every direction and even far from the center.

NGC 1275 is a still more remarkable galaxy than M 87 or M 82 and exhibits some of the more extraordinary characteristics of both (Fig. 9.5). The first structural peculiarities were noted as early as 1931 in the optical observations of E. P. Hubble. Subsequently C. K. Seyfert included it in his group of peculiar galaxies. In 1954 it was shown to be the site of one of the most intense celestial radio sources (Perseus A), and three years later R. Minkowski proposed to explain the observed phenomena as the collision of two galaxies. After the discovery of the explosion in the nucleus of M 82, Sandage and Geoffrey and Margaret Burbidge put forward the hypothesis that this galaxy was also the site of a similar explosion, expanding in every direction instead of along two preferential cones like that of M 82. In early 1970 Lynds, using narrow-band pass filters, obtained an impressive series of photographs, a comparison of which showed that the diverse parts, endowed with various motions toward or away from us, form structures that are completely different in appearance. In particular, the photograph showing the displacement of the Hα line of hydrogen (due only to the radial velocity of the whole galaxy), which records the image of the explosion only in the plane perpendicular to our line of sight, shows a branching structure extremely reminiscent of that of the Crab nebula (which is, as we have seen, an object completely different from this one), abounding with filaments extending to a distance at least four times greater than any yet recorded. These observations confirm the explosion hypothesis and suggest that matter is expanding equally in every direction. Thus the case of NGC 1275 seems to be similar to that of M 82, with the difference that here, perhaps because of the spherical symmetry of the object, the matter expelled is not confined to certain more dense, preferential zones. On the other hand, NGC 1275 is also similar to M 87, since observations carried out in the winter of 1971 from the satellite *Uhuru* have led to the discovery that it is a very powerful emitter of X rays—three times more intense than M 87, the only extragalactic X ray source previously known.

Fig. 9.5 The galaxy NGC 1275, also known as the radio source Perseus A: *above,* NGC 1275 in a normal photograph, which already shows its unusual appearance. (*Mount Wilson and Palomar Observatories*) *Below,* NGC 1275 photographed by Lynds at Kitt Peak in light of the Hα line, allowing for the displacement of the line toward the red (due to the motion of the galaxy), so as to record the hydrogen structure in the plane normal to the line of sight passing through the center of the object. The branching appearance suggests an explosion with expansion of the matter issuing from the center in every direction. (*Kitt Peak National Observatory*)

NGC 1275, M 82, M 87, and Seyfert galaxies present diverse aspects of the same reality: an enormous explosion set off in the center of galaxies of various types or perhaps captured in different phases of one and the same evolution.

It is not out of the question that the explosion phenomenon should be repeated many times in the same galaxy. Thus G. R. Burbidge, on the basis of statistical considerations, taking the number of galaxies showing this phenomenon relative to the total of those known, points out that, given the phenomenon lasts hardly a million years, if it occurs only once, the number of cases of this kind that we are now observing is too high. Thus for each Seyfert galaxy, explosions lasting a million years should occur about a hundred times in the course of ten billion years (a reasonable duration for the life of a galaxy). In the case of galaxies that are also intense radio sources, however (such as M 87 or NGC 1275), explosions should occur only ten times in the same time interval. But is it certain, then, that violent events of this type have never occurred, or could not occur, in other galaxies as well? The possibility should not be excluded: in the nuclei of some spiral galaxies extensive regions of ionized gas have been observed that could represent the origin or the remains of explosive zones; moreover, explosions could occur on a reduced scale, escaping observation. Activity of this kind even goes on in the center of our galaxy, as we have seen; it is so moderate, however, that it would pass unnoticed if viewed from outside at a great distance.

One thing at any rate is now certain: that a galaxy may be not only a stellar city—and an immense one—in which everything is preordained and events succeed one another and evolve with the rhythm of daily life, but also a city that contains a powder magazine, the function and mechanism of which are totally unknown to us. When the magazine explodes, we can no longer see the marvelous harmony suggested to us by the star-strewn spiral arms of so many normal galaxies: we have before us a cataclysm so tremendous and violent that, compared with it, the total explosion of our planet would have no more effect than a pistol shot relative to the explosion of an H-bomb.

INFRARED EMISSION

The terrible explosions that convulse the centers of some galaxies are not the only way their nuclei emit energy. It was discovered that the nucleus of our galaxy emits an enormous quantity of infrared radiation; the same thing can happen in other galaxies, but on such a scale that the emission from the Galactic center, which so amazed us with its intensity, becomes, in comparison, quite paltry. So far twelve galaxies are known in whose nuclei an infrared emission has been found. In three of them it is a hundred thousand times greater than that of our galaxy; in five galaxies this emission has been shown to be a hundred times greater than the usual energy emission of an entire normal galaxy. Between these cases and that of our galaxy there are certainly other, intermediate ones, such as that of the companion of M 51, which radiates in the infrared nearly twenty times more than the Galactic nucleus.

The energy emitted in this way, uniform and continuous, is as enormous as that of the explosions, and its origin is equally mysterious. Thus far no adequate explanations have been found, within the limits of known physics, for the sources responsible for such an emission. And surely many more surprises are in store for us. For example, the twelve galaxies with an infrared nucleus have been observed only at a wavelength of 10 micrometers. But, as observations of the center of our galaxy have shown, there may be an emission in another region of the infrared spectrum, such as the longer wavelength of 100 micrometers, not accompanied by an observable flux at 10 micrometers or even by the emission of light. Perhaps, therefore, there exist true infrared galaxies, that is, galaxies emitting exclusively in the far infrared, that are very intense or nearby but which we are unable to detect because our eyes or photographic plates are not sensitive to the kind of radiation they are sending us.

GROUPS OF GALAXIES

The violent explosions within some galaxies we have been observing are not the only wonder in the extragalactic world. As we continue to explore even vaster regions of the heavens, it will not be long before we discover still other unusual entities. The first thing to strike us will be the existence of strangely shaped galaxies that at least to all appearances do not fit into the categories of elliptical, spiral, or irregular galaxies to which the majority of the other galaxies belong (Fig. 9.6). Many have bizarre shapes; others,

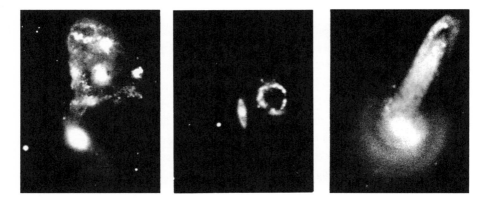

Fig. 9.6   Three galaxies of entirely unusual appearance. *Right,* an example, perhaps, of the interaction of two galaxies, differently oriented in space, so that they are seen by us at different angles. These three photographs are taken from the *Atlas of Peculiar Galaxies,* published by H. C. Arp, which contains 338 of these unusual cases. (*Mount Wilson and Palomar Observatories*)

Fig. 9.7   NGC 5432 and NGC 5435, a pair of spiral galaxies linked by bridges, photographed with the 120 inch Lick telescope. (*Lick Observatory*)

though they are for the most part fairly normal in appearance, present extremely strange peculiarities: they are accompanied by thin or diffuse tails or even by other galaxies of lesser dimensions located in one or more of the spiral arms. Some have adjacent or concentric rings associated with them; others show enormous jets of matter, much larger than the jet observed in M 87. It is not unusual to observe, in the vicinity of peculiar galaxies, matter presumably expelled from the nucleus, and cases are even known in which the very center of the galaxy seems to be about to split in two.

Some of these cases should perhaps be linked with the explosive phenomena that we have already observed; they may represent the beginnings of these phenomena or illustrate their consequences. Perhaps they show both, and if it is true that the explosions are recurrent, these strange galaxies, in ferment rather than split in two, reflect a succession of terrible events and their catastrophic effects. Other cases, however, could be connected to another strange category of galaxies: that of double or even multiple galaxies (see Figs. 9.7, 9.8).

One might think that beneath appearances of this kind there lies nothing really extraordinary and that two galaxies can appear close together simply for reasons of perspective. But this is not so. Such juxtapositions are real, as is shown by the component galaxies' very structure, which is appreciably different from normal: their arms, if we are dealing with spirals, are elongated and distorted by the reciprocal influence of the masses or the magnetic fields concerned. Often, even when the galaxies are fairly distant from one another, photographs have managed to show luminous filaments connecting one with the other across the intervening space. These are actual intergalactic bridges, which are no more than 10,000 light-years wide but often reach lengths of 200,000 light-years. Spectroscopic study, carried out by Zwicky with the 200-inch telescope of Mount Palomar, has shown that they are essentially composed of stars, though sometimes they also contain emission nebulae. In other words, their composition is that of the galaxies themselves (see Fig. 9.9, *left* and *center*).

What do these groups of galaxies represent, and how are the bridges that link them formed? At first it was thought to be a question of colliding galaxies, which met by chance as they traveled separately in space. Today this hypothesis is not favored and it would seem rather that the components of a multiple system are objects formed contemporaneously. Thus it would

Fig. 9.8 The Stefan quintet, a group of five galaxies, already known by the end of the last century. According to radio observations made at Nançay in 1970, one of these galaxies (NGC 7320) is much closer to us than the others and appears in the group only for reasons of perspective. (*Lick Observatory*)

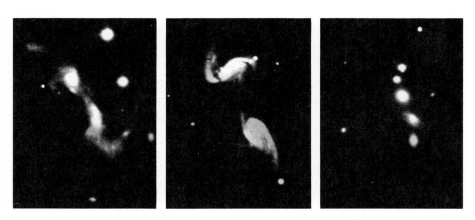

Fig. 9.9 *Left* and *center,* galaxies linked by bridges and appreciably deformed by their mutual attraction. *Right,* a chain of galaxies: the alignment is not due to perspective, as is shown, among other things, by the general background brightness linking one galaxy with another and enveloping the whole chain. Several alignments similar to this one are known today, but this is still one of the most conspicuous. (*H. Arp, Mount Wilson and Palomar Observatories*)

be a question not of "strangers" but of actual "twins." In that case, the formation of bridges would be caused by a tidal effect, by means of resonance phenomena.

The Russian astronomer B. Vorontsov-Velyaminov is of a different opinion, however. According to him, the intergalactic bridges were formed not by a tidal effect, but as a consequence of magnetic fields present in the interacting galaxies.

In addition to double galaxies and small groups (such as the interesting and famous Stefan quintet of figure 9.8), there exist actual chains of galaxies, in which the components are more or less aligned (Fig. 9.9, right). These strange alignments are as mysterious as the groupings we have just observed.

We are almost certainly dealing with appearances that reflect various types or various stages in the evolution of galaxies. For even galaxies, like everything else in the universe, must be subject to evolution.

On the other hand, the small groups of galaxies are only a minute specimen of something that occurs on a much vaster scale. If we explore space systematically, in search of galaxies, we shall soon realize that isolated galaxies are a minority. The majority of galaxies are gathered in groups— not only in those small clusters of five or six that we have just encountered but also in much vaster and more numerous swarms, composed of tens or hundreds of galaxies. As early as 1933 Shapley had counted twenty-five groups of galaxies. As the power of telescopes, especially those of wide field, increased, the number of groups discovered and the number of the galaxies in each group steadily increased. At the same time we greatly increased our knowledge of the most interesting of these groups, which today we do not hesitate to call clusters of galaxies.

The nearest, and hence the best-known, cluster is the one visible in the constellation Virgo. It is about 50 million light-years away and has a mean diameter of 7 million light-years. In the central zone is found the giant elliptical galaxy M 87, while all around there shine at least a thousand other galaxies: the most brilliant are elliptical or spiral; the faintest are dwarf ellipticals like those in the constellations Sculptor, Leo, etc. (which are better known because they are closer to us).

Much more distant, but even richer, is the cluster visible in the direction of the constellation Coma Berenices (Fig. 9.10). There the astronomer G.

Fig. 9.10   The cluster of galaxies in Coma Berenices, in a wide-field photograph taken with the Schmidt telescope of Mount Palomar Observatory. If one looks closely at the numerous white points in the central area of this photograph, one will note, from their slightly elongated shape, that they are not stars but galaxies. (*F. Bertola*)

Abell has observed 1,656 galaxies; he estimates, however, that their total number must reach at least 3,000. In this cluster, 300 million light-years away, the galaxies must be much closer to one another than those in the Virgo cluster.

In fact, even though the diameter of the cluster is 10 million light-years, the clusters are at least three times as numerous—without taking into account the fact that in the Virgo cluster we also observed dwarf galaxies. If these are present in the same ratio in the Coma cluster, the total number of galaxies in the latter must amount to tens of thousands.

The study of a certain number of clusters has shown that some are fairly regular and symmetrical, others less so. The galaxies forming them are of all types, even if the proportions of the various types may vary from one cluster to the next (Fig. 9.11). Often, at the center of a regular cluster, there are one or two elliptical galaxies; among these we sometimes find a so-called D galaxy. Galaxies of this type (so termed according to a recent detailed classification formulated by Morgan) often have double or multiple nuclei and emit radio waves of considerable intensity. When a cluster of galaxies contains such a galaxy, it is bigger and brighter than any other member of that cluster and in fact occupies the center of the cluster.

One might think that just as stars can be grouped in clusters (which in turn form those much richer and more complex systems, the galaxies), so also clusters of galaxies could constitute the basic elements of even vaster and more numerous agglomerations. This point is still the subject of controversy. Some astronomers (such as Abell, van der Bergh, and de Vaucouleurs) hold that clusters of galaxies are composed of a number of smaller groups (such as the Galaxy and the Magellanic Clouds, M 31 and its two companion galaxies, and so on) and in turn constitute a unit that forms much vaster and more abundant groups: superclusters. In that case the Galaxy and its neighbors, forming the so-called "local system of galaxies," would be part of a system of clusters having as its center the Virgo cluster.

Other astronomers (such as Zwicky) vehemently deny the existence of superclusters of galaxies, insisting that these are only clusters of galaxies that are more or less extensive and more or less abundant, since there exist, as well, many isolated galaxies, that is, galaxies that do not belong to any cluster.

Fig. 9.11   The central part of the cluster of galaxies in the constellation Hercules. Here one can see clearly (better than in the cluster of Coma Berenices) that the clusters are composed of a population of galaxies of all types and even exhibit examples of interacting pairs, including two spiral galaxies. (*Mount Wilson and Palomar Observatories*)

It may be that the controversy is due, at least in part, to the very definitions of cluster and supercluster assigned by the various authors. Certainly, the incompleteness and fragmentary nature of the observations made so far have also constituted a great obstacle.

This gap, however, has been filled recently, largely by Zwicky himself, who succeeded between 1961 and 1968 in compiling and publishing (in collaboration with other astronomers) the most extensive catalog of galaxies and clusters of galaxies ever made. Even though this catalog does not include the Southern Hemisphere, it covers the whole northern sky (from the Pole to 3° S) and contains an enormous number of galaxies. The isolated, or larger, galaxies number 31,350, whereas the number of clusters amounts to 9,700. Of course some of these clusters are composed of only a few tens of galaxies, but there are others that contain more than a thousand: one actually contains 3,755 galaxies, another 5,130. Assuming that, on the average, each cluster comprises at least 200 galaxies, it follows that at least 2 million galaxies have been counted.

But an average galaxy contains about fifty billion stars. We have, therefore, direct proof of the existence of at least a hundred million billion stars. If every star were the center of a system of lesser bodies (as is true of the sun), how many planets, satellites, comets, and asteroids must there be in all of this portion of the explored universe? The number cannot be specified, because we still have no idea of the percentages of planetary systems relative to the number of stars. We can only state that the number must be enormous when we reflect that, even if there were only one planetary system like ours in each galaxy, we should have to admit the existence of at least eighty million bodies like the planets or satellites and fifty million billion smaller bodies like the asteroids and comets. The hypothesis of only one planetary system per galaxy, however, is unlikely, for while visiting the stars near the sun we already ascertained the existence of other planets, and we noted that, in the case of the more distant stars, we were unable to discern planets only because our means of observation were limited.

Thus the number of the planets, satellites, and comets whose existence we have posited should be multiplied by an extremely high factor: at least a million, perhaps a thousand million.

These are figures enough to boggle the mind of anyone who grasps their extent and significance. It is difficult to comprehend their value, and if one succeeds, it is only for a moment, for the need to forget them is stronger than the desire to contemplate them. We feel like speleologists who return suddenly to the light of day after spending hours or days in the darkness of caves: they have had to shut their eyes to block out the torturous light of the sun. Later, when little by little they open their eyes and look at the landscape, they realize that although the light was violent and cruel, what it now illuminates is beautiful. And I am not only thinking of the usual flowery meadow or romantic landscape, for even the most arid desert, the most desolate ice floe, or the most squalid surroundings in the most dismal slums are beautiful compared to darkness—beautiful because of their colors (which we so often forget simply because we are so accustomed to them), beautiful because of the variety of their shapes, and beautiful and interesting because, whatever the sight, it contains an enormous number of things to discover.

And we too shall yet discover so many things, if we manage to cope with these huge numbers that enlighten us as to the structure and nature of the universe, which is continuously unfolding before our eyes.

No sooner do we resume our journey, in fact, when we discover that the figures just quoted are the product of a computation that is only approximate and certainly incomplete in its accounting. In fact, it is based on Zwicky's catalog, which does not enumerate all existing galaxies, nor even all those observed thus far. In reality, it leaves out not only all the galaxies in the southern sky (as we already know) but also all those fainter than apparent magnitude 15.7, observable only with instruments more powerful than the two Schmidt telescopes at Mount Palomar with which Zwicky carried out his survey. Many other clusters are also known, which are faint because of their enormous distance; these are accessible with larger telescopes, such as the great 200-inch reflector at Mount Palomar. Unfortunately this reflector has a much more restricted field than the Schmidt telescopes, and it would require thousands of years to photograph all the sky visible from Mount Palomar, which the Schmidt telescope photographed in only seven years. Nevertheless, some abundant and very distant clusters in various parts of the sky have been discovered with this reflector, which leads us to believe that, even if we continue to move far beyond the limits of Zwicky's catalog, we shall find galaxies and clusters springing up everywhere in space.

But where, then, is the limit to all this, if indeed there is a limit? If we want to try to discover it, all we can do is go on; when one knows nothing, the best way to find out where a road leads is to follow it. But we cannot advance immediately. The question that we have just set ourselves is one of those which have tormented man since he acquired the use of reason, and if we wish to try to find an answer that is, if not definitive, at least satisfying and reasonable, we shall need the greatest possible store of knowledge. Thus the subject of the limits and structure of the universe will be the last that we shall tackle, and we shall find that we have done well to wait, for we are about to encounter still more phenomena and objects that will be of great assistance to us in this final discussion. Therefore, we shall stay where we are for a moment longer, to search not the galaxies themselves but *between* them.

INTERGALACTIC SPACE

When, after leaving the Galaxy and the Magellanic Clouds, we found ourselves for the first time in intergalactic space, the first impression that struck us was that of the enormous cosmic void. Subsequently, after traveling a distance of over 2 million light-years, we reached M 31, another huge aggregate of matter and stars—another galaxy. Beyond it we encountered others: so numerous and often so enormous, but all, like the previous ones, very distant from one another. We thus discovered the world of galaxies, but there also remained, between one and the next, those spaces completely devoid of matter or any kind of heavenly body—vacant and desolate spaces. Many astronomers (especially Zwicky) did not find this picture very convincing; they could not accept the idea of an intergalactic space that is completely empty. Their research in this direction has led to some very interesting results.

The first indication that there is other matter between our group of galaxies and that of Andromeda was given by the discovery of the dwarf galaxies. The objection might be raised that, however small these aggregations might be, they are still galaxies, and in fact their discovery only indicates that the number of galaxies is greater than it seemed at first and the volume of empty space between the galaxies is slightly less. But if faint galaxies had escaped our notice, was it not possible that, a fortiori, smaller bodies, such as globular clusters or single, scattered stars, might also remain unnoticed? Subsequent research has given a positive reply.

In 1955 Abell published a list of thirteen globular clusters discovered with the big Schmidt telescope at Mount Palomar. Subsequent observations of some of these showed that they are extremely distant, far beyond the reaches of the Galaxy. Thus in intergalactic space we may find not only small galaxies but also minor stellar aggregations such as globular clusters.

And we may also find, scattered here and there, isolated stars. Their existence has been proved by the discovery of blue stars of high intrinsic luminosity but faint apparent magnitude, observed at elevated galactic latitudes. Obviously, if such stars are, as is demonstrated by their color, intrinsically highly luminous but appear so faint, this can only be due to their enormous distance. When we recall that our galaxy is an extremely flattened system and that we are located near its equatorial plane, it is clear that, when they are seen in a direction away from this plane, their great distance places them beyond its limits. Of course, even though they are extragalactic stars, they are still relatively close to the Galaxy. But nothing forces us to exclude the existence of other, more distant stars that we do not see, simply because their apparent brightness is so reduced by their distance that none of our existing telescopes is able to reveal them.

Moreover, their presence is suggested not only by this extrapolation but also by another extremely interesting discovery: that of the intergalactic bridges. The existence of a bridge of stars between two galaxies is, in itself, the best proof of the presence of stars in intergalactic space. One might think that this is a very special case, in which the idea of "intergalactic space" loses much of its original meaning; after all, when we speak of intergalactic space, we instinctively tend to think of the space between the Galaxy and M 31 rather than of that between the Galaxy and the Magellanic Clouds. With the passage of time, however, the two galaxies linked by the bridge of stars move steadily farther away from each other, and most of the stars of the bridge remain isolated from one another and are dispersed in intergalactic space. Thus, since there exist bridges between galaxies, we must also find scattered stars resulting from the breakup of bridges that once existed between galaxies now extremely distant from one another.

Since there are clusters and stars in intergalactic space, one might suppose that there is also gas and dust. According to Zwicky there are proofs of this. The most convincing is based on computations of the galaxies in each cluster. Zwicky has proved that when the fainter (and thus more distant) galaxies are counted, there comes a point at which the sky background around the cluster appears more abundant than the central zone of the cluster itself. This shows the presence of a certain amount of dust within the cluster, capable of obscuring—or even rendering invisible—many of the galaxies located in that direction.

The presence of intergalactic dust also seemed to be confirmed—sensationally and unexpectedly—by the German astronomer C. Hoffmeister some fifteen years ago, in the course of research on RR Lyrae type variables far from the equatorial plane of the Galaxy. In order to show possible interstellar absorption—within our own galaxy, that is—he had decided to study the distribution of the galaxies present in the same field. Examining a region in the constellation Microscopium, he detected the existence of an irregular zone of about 20 square degrees that contained only five galaxies, instead of the twenty to thirty he expected. In the same zone, however, the abundance of RR Lyrae stars remained normal. This fact showed that there must be a dark cloud between these stars and the very distant galaxies in the background. Given that the RR Lyrae stars are distributed within the Galaxy to great distances, the fact that the cloud did not affect their number showed that it should be assumed to lie beyond them, that is, outside the Galaxy. Bearing in mind the Galaxy's considerable apparent dimensions, the conclusion was reached that even if the dark cloud were very close to the Galaxy, like the Magellanic Clouds, it must have a diameter of at least 40,000 light-years.

Have we thus discovered a whole black galaxy, perhaps composed of black, completely dead stars? No, we do not wish to make any such claim, and indeed we can be certain that if such a galaxy existed, we should not be able to discover it, at least in this way, because it would not cause the slightest absorption: light from bodies in the background would reach us all the same, passing between one dark star and the next. Even if the latter were very numerous, they would still occupy a very small volume relative to the vacuum that separates them.

The case is very different for dust grains, which are much smaller than stars but far more numerous. If they form a cloud of sufficient thickness, the absorption becomes appreciable.

Even the immense dark cloud indicated by Hoffmeister, if it indeed exists (which recent observations have rendered doubtful), must be composed essentially of dust for, in any case, it cannot contain gas since none is revealed by radio telescopes.

Clouds of intergalactic gas seem to exist nonetheless. According to research by Sargent and L. Searle, published in late 1970, two objects that Zwicky listed as compact galaxies are in fact dense clouds of neutral hydrogen isolated in intergalactic space, 24 million light-years away from us. Their diameters are less than 1,000 light-years—825 and 650, respectively; the mass of the second, according to radio observations, amounts to 230 million times that of the sun. The spectra show the lines typical of gaseous nebulae in a state of high excitation. Thus they resemble the Tarantula nebula, which we encountered in the Large Magellanic Cloud, and like it must contain a great number of O-type stars at a very high temperature, which are responsible for the excitation. The second cloud is thought to contain from a thousand to a hundred thousand such stars. Specific infrared observations carried out by G. Neugebauer have enabled us to exclude any appreciable contribution to the luminosity by red stars, which must constitute at most a tiny minority. We are dealing, then, with two isolated zones of gas, containing a very large number of stars, of great mass and extremely young. Thus we have here, isolated in intergalactic space, a new category of objects, which previously we knew only as components of a galaxy. They are, it is true, exceptionally large objects, but their presence does not exclude—but rather induces us to acknowledge—the existence of other similar objects, smaller and fainter, which elude us only because we cannot observe them.

Dwarf galaxies, compact galaxies, clusters, isolated stars, gas clouds, and perhaps even dust clouds of varying densities—these are what a more thorough search has led us to discover in that intergalactic space which, at first glance, appeared completely empty. Of course this does not mean that the average density of matter in intergalactic space is high or even comparable to that within a galaxy. On the contrary! Even though we cannot yet form a clear idea, there are grounds for believing that it is relatively slight.

The bodies that escape us are only the fainter or smaller ones—in any case, those composed of a smaller amount of matter—and, moreover, the volume occupied by the space between the galaxies is enormous, even relative to that occupied by all the galaxies put together.

The question of the mean density of space is one of the most important subjects in modern astrophysics, and it continues to be widely discussed. But even if we were to discover large quantities of matter, hidden masses, eluding us in forms that have only begun to be detected in recent years, it is most unlikely that they would be sufficient to raise appreciably the very low mean density of intergalactic space. Thus although we have discovered that between the Galaxy and M 31 there are a few dwarf galaxies, various clusters, and an indeterminate number of stars, it is reasonable to conclude that, on account of their smallness and the immense distances separating them, we could not see these objects with the naked eye unless we were in their vicinity. Thus the true picture of intergalactic space remains as it was originally: a vast dark abyss, in which two or three faint specks stand out— the brightest and nearest galaxies.

Let us imagine that we are on a planet belonging to a star between the Galaxy and the Andromeda nebula. At sunset, if no satellite lights up the night, a dark, empty sky will descend on us, rendering the landscape around us dark and empty. It is the same view we saw as we left the Galaxy, but now we are experiencing it with our feet planted on solid ground, ground that we cannot see, just as we do not see our feet pressing into it, or our hands, which we hold before our wide-open eyes, trying in vain to discern them. This is what the space around us and the sky above us have become with nightfall! The blue vault of the sky, transparent, ethereal, has become increasingly dark, gloomy, and finally black, impenetrable, hard as the vault of a cave.

When we left the Galaxy, this was a fantastic vision that we thought available only to hypothetical astronauts who had traveled between two widely separated galaxies, where they encountered no stars. Now we know, however, that there are such stars, and perhaps there are beings near them who have never known the beauty of the starry sky.

THE GALAXIES MOVE

So far we have spoken of the galaxies as if they were all motionless in space. It is now well-known that all bodies are in motion. As Heraclitus said 2,500 years ago, everything flows, and subsequent discoveries have further confirmed this truth, proclaimed by the ancient philosophers. The earth turns on its axis; the moon revolves around the earth, and the moon and earth around the sun; planets and satellites rotate and move like the earth and the moon; the whole solar system moves in space; stars pulsate, nebulae expand, star clusters move and break up; the entire Galaxy rotates, and all other galaxies rotate or expand. The whole universe, from the atom to man, from the planets to the Galaxy, is constantly in motion. Everything passes, changes, becomes. Thus there was no reason to believe that the galaxies should be any exception or that, although they are composed of an enormous number of bodies in motion, they should be, as a whole, static entities, completely immobile in space.

On the other hand, now that we know how great the distance is between the galaxies, we may certainly abandon all hope of observing their apparent displacement on the background sky, even in the case of those closest to us and even comparing photographs taken at intervals of tens or (when we are able) hundreds of years.

There is, however, a method for discovering at least that part of their motion which takes place along our line of sight. It is the same method that has already been used so widely for the stars: the observation of the displacement of the spectral lines due to the Doppler effect, which gives the actual radial velocity, that is, the velocity of the body along the straight line that joins the point occupied by the body to the position of the observer.

The first galaxy for which the radial velocity was successfully measured was M 31, for which V. M. Slipher found, in 1912, a velocity of approach of 300 kilometers per second. Actually the motion does not correspond solely to the displacement of M 31 but to the combination of its motion with our own—that is, with that of the sun in our galaxy. In other words, we are dealing with a relative motion. More recent measurements, which take this effect into account (that is, they are based on the assumption that we are at rest), have led to the conclusion that M 31 is moving toward us at a velocity of 35 kilometers per second. After 1912 Slipher persevered in the difficult task of obtaining spectra of galaxies;[1] thus by 1925 the radial velocities of forty-five galaxies were known. Even in this early material a baffling result began to emerge: nearly all the galaxies observed were found to be moving away from us. Among the few exceptions was M 31, the galaxy Slipher had measured first, since it is the brightest.

Meanwhile, the distances of a certain number of galaxies began to be known. In relating the velocities of recession to the distances, Hubble made a sensational discovery. He found that all the galaxies (with a few exceptions, among those closest to us) are moving away from us, and the greater their distance, the higher their velocity of recession. In other words, all the spectral lines of the galaxies appear shifted toward the red, the more so the farther away is the corresponding galaxy (Fig. 10.1). The result can be expressed in the following formula, as important as it is simple: $V = Hr$, where V is the velocity of recession, r the distance, and H a number known as the Hubble constant. In practice one obtains different values of H for various galaxies and then calculates the average. Hubble's law, discovered in 1929, has been confirmed by subsequent observations.

In 1928 Humason undertook a spectroscopic survey of the greatest possible number of galaxies, using mainly the 100-inch telescope at Mount Wilson (then the largest in the world). Thanks to this enormous work, later resumed with the 200-inch telescope at Mount Palomar and continued for more than twenty years, the red shifts of a thousand galaxies are known today.

It was more difficult, however, to determine the precise distances. But the methods used in this kind of research have also undergone constant improvement, and today the distances of many galaxies are known fairly well, though for many fewer galaxies than those for which it has been possible to determine the red shift. It is thus possible to derive a good value for the Hubble constant. According to the most recent measurement, $H = 55$ kilometers per second per million parsecs. The interpretation of Hubble's law has been debated at length. If we assume that the shift of the spectrum toward the red is due to the Doppler effect, we must conclude that all the

---

[1] The work is difficult for a practical reason, due to the fact that the galaxies appear not only very faint but also extended, and their spectra can be obtained only with the most powerful telescopes and with exposures of many hours.

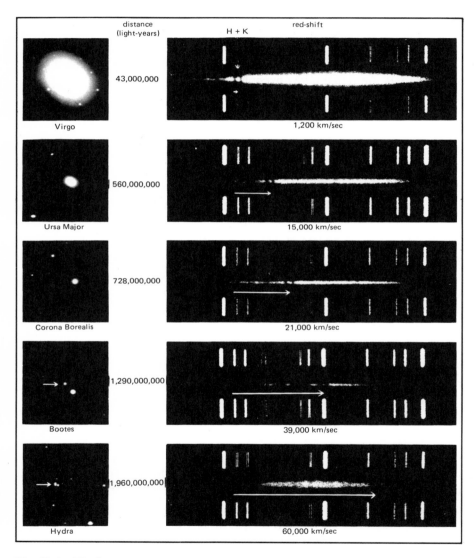

Fig. 10.1   Displacements toward the red (red shifts) observed in the spectra of galaxies increasingly distant from us. Note the spectral lines shifted more and more to the right. (The arrows indicate the displacement of the H and K lines of calcium.) A photograph of the corresponding galaxy accompanies each spectrum. (*Mount Wilson and Palomar Observatories*)

galaxies really are receding from us, at velocities that increase the farther they are from our galaxy (Fig. 10.2).[2] Thus one might be inclined to think that the Galaxy occupies a very special and exceptional position relative to the other galaxies, which is to say, in the whole universe. That is a relic of geocentrism that should at least make us suspicious. At various epochs man has placed at the center of the universe first his own territory, then his planet, his sun, and now, finally, his galaxy. But what is so special about our galaxy that all the others obey it, if only in the sense of fleeing from it? And what happened, or is happening, here within our galaxy to set in motion a mechanism so colossal that the tremendous explosions we have seen in individual galaxies pale beside it? Undoubtedly, either there is something wrong with this interpretation of Hubble's law or we are faced with a vast new mystery. In fact, as we shall see later, in such an expanding universe one would seem to be at the center wherever one is.

Perhaps the red shift of the spectral lines is due to some physical phenomenon other than the Doppler effect, a phenomenon of which we are unaware because we are applying the laws of physics to a time scale and a scale of distance completely different from those of the laboratories in which they were discovered. Many of the laws discovered on the scale of human sensations are not valid on the atomic scale, which is ten million million times smaller. Why, then, should they hold on the extragalactic scale, which is ten million million times larger? Perhaps Hubble's law can be interpreted on the basis of the Doppler effect, but by adding further considerations and hypotheses on the nature of the universe. It is possible, and we shall see in the end how far we can go.

For the moment let us simply reject the absurd hypothesis that we live in a privileged galaxy, but let us not forget that Hubble's law is a positive fact, for it expresses a phenomenon observed in reality: the proportionality between the red shift of the spectra and the distances of the galaxies from us. Even if accepted solely in these terms, therefore, it becomes a very powerful means of obtaining the distances of the galaxies. Indeed, on the basis of the law, the distance of every galaxy can be immediately found by dividing the value of the red shift of its spectrum by the Hubble constant, which is already obtained by means of the galaxies whose distances are known.

[2] The few velocities of approach are explained as being due to local motions.

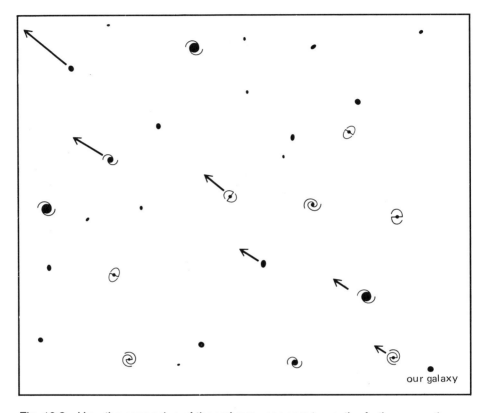

our galaxy

Fig. 10.2   How the expansion of the universe appears to us: the farther away the galaxies are, the more rapidly they seem to recede from us. (*Adapted from P. W. Hodge*)

Therefore, from now on we find out the distances of even the most distant extragalactic objects merely by managing to determine their spectra. And we have thus succeeded in discovering that many of the galaxies that we have already observed and photographed are hundreds of millions of light-years away.

By this means we have measured the distances of single galaxies and of clusters of galaxies up to about 8,000 million light-years. We have enlarged our horizon—or rather we have realized the immensity of what we could already see. This time we have not simply rediscovered our physical insignificance (which we have already known for some time) but the possibilities of our intellect and of the methods of observation that we have succeeded in inventing, which permit us to advance so far. Now we must explore more thoroughly what this new horizon discloses and its limits before we try to advance farther still.

THE QUASARS

If "quasar," the name of the last object with which we shall preoccupy ourselves, should smack of science fiction to newcomers, I hasten to assure them that perhaps they are not wrong. We are now moving beyond those concepts and the knowledge that has become familiar to us in the first half of this century, and we are entering a world in which science and fantasy intertwine and alternate in the new visions that arise from the direct exploration of the cosmos and from the daring (but realistic) interpretations that scientists are trying to outline.

The quasars are the most mysterious of recently discovered extraordinary objects, entities with characteristics so unusual that man's imagination would never have ventured to conceive them with any conviction that they might bear even a semblance to reality. They do, however, exist, and observations prove it.

Their discovery goes back to late 1962, when, through comparisons of various phenomena already observed a few years earlier with some new observations, a new class of objects emerged, with incredible properties—so incredible that, at first, when only a few were known (and those not well observed), many astronomers still would not accept their reality, or at least all its aspects. Today more than five hundred are known, many of which have been followed and thoroughly studied in the past fifteen years, and even if their nature remains shrouded in mystery, at least there can no longer be any doubts about the reality of the phenomena they exhibit. But let us get to the facts.

The first characteristic discovered was a very intense and highly concentrated radio emission. When this emission was precisely located on the celestial sphere, it was found to come from objects of stellar appearance. When they were studied by traditional optical methods, strange anomalies soon appeared, and subsequent research showed that, despite their appearance, they are definitely not stars. The name *quasar* (an abbreviation for quasi-stellar radio source) was then coined, but it was not until late 1970 that the name was definitively adopted, with a modification that we shall soon come across. The brightest quasars were photographed and observed spectroscopically at Mount Palomar Observatory, with the world's largest telescope. The spectra showed a continuous background (later, in some cases, absorption lines were also seen) and a few emission lines that were not at first identified with those previously observed in heavenly bodies. Subsequently, however, it was found that they corresponded to known lines (the principal lines of the Balmer series for hydrogen, and some others), greatly shifted toward the red—to the point where it was impossible to recognize them right away. This phenomenon could be interpreted in three ways.

One was based on the Einstein effect. According to the theory of relativity, if a luminous source lies in a strong gravitational field, the light undergoes a displacement toward longer wavelengths (that is, toward the red); the higher the value of the mass of the body causing it, the greater the displacement. One might conclude, then, that quasars are very heavy stars, but this interpretation runs up against various problems. First of all, it is calculated that the masses required to explain the observed displacement must be enormous, particularly relative to the volume in which the mass must be contained. Then, supposing that the heaviest objects are also the brightest (as is generally the case with stars) and assuming that they are all at the same distance (the problem introduced by the nonvalidity of this second hypothesis can be overcome by statistical considerations), we should observe greater red shifts in the brighter objects and lesser shifts in the fainter ones. In reality, however, the opposite is observed.

But there is more. If there were bodies as heavy as this and little larger than stars, matter would be subjected to an enormous pressure, which would cause an appreciable broadening of the spectral lines; this, however, is not observed. Furthermore, the neighboring stars would undergo perturbations large enough to be observable (at least, in the more favorable cases) with our present means of observation. Thus this first interpretation should certainly be discounted.

A second hypothesis supposes that the red shift can be interpreted as a Doppler effect and that we are dealing, therefore, with stars of extremely high velocity, relatively close to us. In that case, however, we note that the observed red shifts give velocities of the order of magnitude of some tens of thousands of kilometers per second, and unless the objects are moving only along our line of sight, there should be a transverse component of the motion corresponding to an appreciable displacement of the object on the celestial sphere, which could be easily measured by comparing observations a few tens of years apart. Such comparisons, made by using old photographs on which some quasars had been recorded, have had negative results. Since it is inconceivable that the motion of all the observed quasars should take place exactly along the direct line of sight from the earth, and, moreover, since there would still be no explanation for the fact that all are velocities of recession (all the displacements observed are to the red), this second hypothesis must also be rejected.

In 1966 this second hypothesis was taken up again in another form by Terrell and supported by F. Hoyle and G. R. Burbidge. According to these scientists the quasars actually move at the very high velocities observed; they are not stars or bodies that belong in any way to the Galaxy, however, but fragments of extragalactic matter hurled into space by colossal explosions. They are less than some 30 million light-years away. In fact the authors of this interpretation even suggest as a possible site of the explosion the galaxy NGC 5128.

This theory has also attracted much criticism. The most obvious objection is still that no quasars are observed with the spectrum displaced toward the blue, that is, approaching us. If a body outside the Galaxy had exploded and had hurled lumps of matter in every direction, some should also be moving toward the Galaxy. But there is also a second objection. L. Woltjer and G. Setti have calculated that, taking into account the actual kinetic energies of the quasars (that is, the energies associated with their velocities) and considering that the sum of these energies must have been provided by the explosion, the amount of energy that the original explosion must have had is absolutely inconceivable.

The third interpretation of the red shift is that it is of the same nature as that observed in the galaxies. In this case, if we apply Hubble's law, the distances of the quasars should be proportional to their red shifts and can thus be derived when the latter are known. On the basis of this interpretation, which remains the only one that is possible, the quasars are found to be extremely distant from us—most of them over a billion light-years.

The most distant objects so far observed by man are, in fact, two quasars: OQ 172 and OH 471. Their red shifts are $z = 3.53$ and $z = 3.40$, respectively. Thus their distances from us would be on the order of 15 billion light-years. Of course, most of the other quasars are not so far away, but they are still generally some billions of light-years distant. If their distances are so great, a simple calculation shows that their brightness must be enormous. Thus 3C 273 (Fig. 10.3), whose apparent magnitude is 12.5, must have—given its distance—an absolute magnitude $M = -25$, which corresponds to a trillion times the brightness of the sun. Such a brightness is greater than that of a whole galaxy (even the largest), which is nevertheless an immense aggregate of matter and of hundreds of thousands of millions of stars. Quasars are, therefore, the most luminous objects known in the universe. So much released energy, emitted not only at optical frequencies but also in the domain of radio waves, is given off within a relatively small volume. Indeed, even when observed with the most powerful telescopes, the quasars continue to appear starlike. In practical terms, it is calculated that they contain masses equal to tens of millions of times that of the sun, within a space of a diameter no greater than some thousands of light-years.

But the surprises are not over. As early as 1963 astrophysicists all over the world were astonished by the news that examination of old plates covering a timespan of about seventy years, on which the quasar 3C 273 happened to appear, revealed appreciable variations in brightness. Subsequently, precise photoelectric observations confirmed small changes at brief intervals, in both this and other quasars (Fig. 10.4).

This was a matter of enormous importance, because it reduced even

Fig. 10.3   3C 273, one of the brightest quasars. This photograph, taken by A. Sandage, clearly shows a "jet" that extends about 150,000 light-years from the brighter nucleus. The jet, an intense source of radio waves, is composed of ionized gases, whose electrons move on paths spiraling within the lines of force of the magnetic fields. The matter was expelled from the nucleus about a million years previously. The light of 3C 273 takes several billions of years to reach the earth. (*Mount Wilson and Palomar Observatories*)

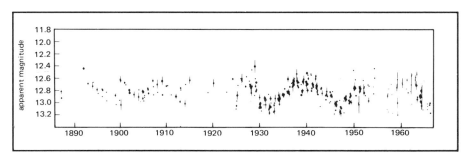

Fig. 10.4   A graph showing the variations in the visual light of 3C 273, which H. J. Smith and his colleagues discovered by examining photographic plates going back over a period of eighty years. Plotted against the date of each exposure are the determinations of the apparent magnitude: since an interval of one magnitude corresponds to an intensity ratio of about 2.5:1, it follows that, during this period of observation, 3C 273 fluctuated up to 50 percent in luminous intensity.

more the dimensions of the quasars—or, at least, of that zone in which most of the optical energy is produced and where the phenomenon of light variation appears. Indeed, if a quasar had a radius of, for example, 1,000 light-years and there occurred for some unknown reason a phenomenon that changed its brightness over its entire surface at the same instant, we should observe the variation in brightness of the part of the surface closest to us at a given moment, and that of the farthest part, the rim, a thousand years later. (Of course, during these thousand years we should observe the same variation coming, successively, from all the intervening zones.) Since the variation can be observed all at once, the quasar (or at least that part of it subject to variation) must be relatively small.

In 1965 W. A. Dent, observing 3C 273, discovered that the radio emission was also variable. Soon thereafter other radio astronomers observed various quasars and discovered that the phenomenon was common to many others. After the extraordinary news of the optical variability, however, this fact was pretty much expected and not so sensational.

A great sensation, however, was caused by the announcement that a quasar (3C 446) had undergone an increase of 3.2 magnitudes from the beginning of October 1965 to the end of June 1966. Unfortunately, no other observations were made during this entire interval, but there is reason to believe that the increase was very rapid, since back in September 1965 an astronomer who had observed it spectroscopically considered it to be at its usual brightness—that is, at minimum. Thus 3C 446, although it is a body billions of times brighter than the sun, was able to increase its brightness fully twenty times, and perhaps almost instantaneously. Of course, after showing such a burst, it has been followed continuously by astronomers, who have been able to ascertain that it continues to vary around the maximum with exceptional irregularity (Fig. 10.5).

Observations of the variability of quasars have continued, disclosing increasingly interesting details. It seems that at least a third of those subject to constant observations have shown phenomena of variability. The variations in light are generally modest (less than a magnitude) and irregular. Bursts such as that shown by 3C 446 have not been observed in other quasars; hence they must be very rare. One thing is certain, at any rate: the phenomenon of variability is real and striking, and, however inexplicable it may be, it forces us to consider quasars as objects that are relatively very small—that is, with a diameter of a few hundred light-years (Fig. 10.6).

The fact that such small objects can produce so much energy and perhaps be so heavy becomes an ever deeper mystery.

In 1965, a few years after quasars were discovered, the astronomer Sandage made another important discovery: he found objects that had optical properties similar to quasars (stellar appearance, excess of ultraviolet light, red shift of the spectral lines) but which, unlike quasars, did not seem to emit radio waves. He called them QSG (quasi-stellar galaxies). These objects were much more numerous than quasars and yet, in view of their optical properties, it could not be denied that they might belong to the same family. By the end of the same year this fact was confirmed by radio astronomers in Bologna who discovered in some of these objects a radio emission, much weaker, however, than that observed in the quasars. By now this weak radio emission has been detected in many of the others. At any rate, M. Schmidt proposed the name *quasar* to denote objects of this type, whether radio emitters or radio-quiet, and it has been officially accepted in this sense by the *Astrophysical Journal* (November 1970), one of the most authoritative astrophysical periodicals in the world.

The number of these objects is thought to be very great. Since they are not revealed by radio emission, discovering them is a long, laborious task, for it is difficult, with rapid methods of observation, to distinguish them from stars. However, on the basis of the number discovered in the zones explored thus far, we can estimate that the total number, down to those of apparent magnitude 22, should be about ten million: ten million objects, as heavy as galaxies and brighter, but pointlike in appearance, hence undoubtedly very much smaller—even smaller, perhaps, than the nuclei of Seyfert galaxies, which they resemble. What is their nature and, above all, what do they represent in the general economy of the universe? We do not know. But we can still try to imagine their structure, at least in outline. Theoretical astrophysicists, at least, are trying to do this, starting with the principal observed properties that are firmly established, which we shall briefly describe.

To recapitulate: A quasar is an object that is stellar in appearance and may or may not be a radio emitter. It shows a continuous spectrum, some-

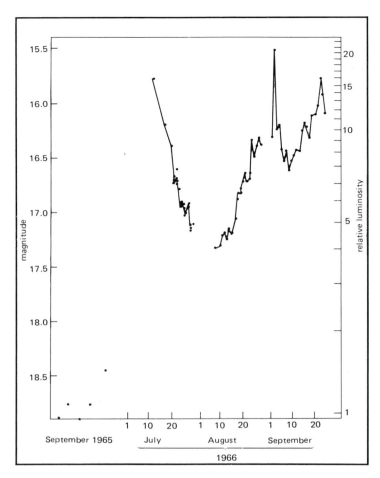

Fig. 10.5   Light curve of 3C 446, a rapidly varying quasar, from observations by T. D. Kinman, E. Lamla, and C. A. Wirtanen.

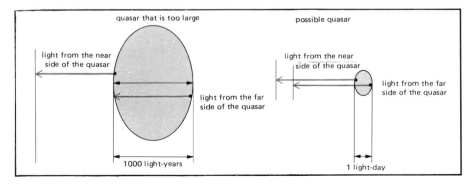

Fig. 10.6   Relation between the dimensions of a quasar and its variability: given a particular period of variability for a luminous object, its dimensions can never exceed the distance traveled by light in that period, for simple optical reasons.

times crossed by absorption lines, and, above all, superimposed on the first, an emission line spectrum, corresponding to that of highly excited nebulae and greatly shifted toward the red; the latter fact suggests quasars are extremely distant extragalactic objects. Their spectrum also shows an excess of ultraviolet and infrared radiation, relative to that of the stars.

For this reason, and because of the enormous difference between it and the line spectra of stars, the spectrum of a quasar, unlike the spectra of galaxies, does not appear similar to stellar spectra. Thus a quasar cannot be an aggregate of stars like a galaxy, even though it may be reduced to a pointlike appearance by its compactness and enormous distance. Finally, the variability of the brightness of quasars confirms their small size without, however, disproving their extragalactic nature, since the nuclei of some Seyfert galaxies (as we have already seen) may also be variable.

From these observations, as well as radio observations (which we have not discussed), it seems possible to conclude that a quasar is composed essentially of three parts. The first is a central body with a diameter of about 2 light-months, that is, about 120 times greater than that of our solar system. Its mass would be at least 100 million times that of the sun, and its absolute visual magnitude -24. The energy that it radiates over the entire optical range, including the ultraviolet and the infrared, is estimated at 10^{46} ergs per second, that is, 10 trillion times more than the energy emitted by the sun in the same optical range.

The central body is surrounded by a much lighter gaseous envelope, which is expanding at a velocity between 1,000 and 10,000 kilometers per second and extends up to 3,000 light-years from the center. This atmosphere, which has a rather irregular and filamentary structure, should have a mass equal to a million times that of the sun. The atmosphere, in turn, is enveloped in a corona of high-energy particles, which has an extremely low density and a radius of some tens of thousands of light-years; this also is not uniform in structure. It is in this part of the quasar that the radio emission originates, and when no emission is recorded, there is no corona. In that case the object is composed solely of the other two parts and belongs to the radio-quiet group discovered by Sandage. Probably this corona does not constitute a stable and permanent part of the quasar but is produced by a rare, extraordinary, violent event, or by several such events, which accelerate the particles to a velocity approaching that of light; the effects of these events may last from a thousand to a million years. This is how a quasar must appear, at least in its basic structure. But no one has yet figured out what it is, how it originated, how it manages to emit the enormous amount of energy observed, and finally, since it cannot remain forever as we see it now, how it will evolve.

Various theories have been formulated concerning this point. Among those which are possible, the one devised by L. Gratton in 1967 is very interesting. Starting with the equations for the mechanical equilibrium of a spherical fluid mass, in the relativistic case, and adopting the hypothesis that the transportation of energy within the central body takes place essentially by convection, Gratton developed a quasi-static model of a quasar that emits energy as it expands.

This model taken with the fact that a quasar, though much smaller than a galaxy, is a billion times brighter than the sun and hundreds of millions times heavier, suggests a bold evolutionary hypothesis. Perhaps, by expanding through a series of equilibrium states, a quasar is transformed in ten or a hundred million years into an immense cloud of gas, dark and cold. At a certain moment, condensations of this cloud would start to form, from which stars would be born. At this point the process of expansion would come to a halt, while the quasar was being transformed into a conglomeration of gas and stars. Thus from the quasar a galaxy is born.

A COSMIC VIEW

When we first ventured beyond the Galaxy and the Magellanic Clouds, and, with our weak vision, peered ignorantly into space, we had the depressing feeling that we were lost in a void. Then, guided toward a luminous piece of fluff, we ended up discovering other galaxies—indeed, the world of galaxies. Now that we have appreciably broadened our knowledge of that world, we should like to repeat the experiment of contemplating it from outside the Galaxy, only this time relying, not on our very limited senses, but on those means which have enabled astronomers to make all the wonderful discoveries with which we have become acquainted.

Let us imagine, then, that we are looking toward these same regions of space with eyes thousands of times more powerful than our own, that is, as if, instead of our own pupils, we had pupils 625 times wider, the size of the telescope at Mount Palomar. What we need, in fact, is to grow somewhat bigger, to reach a height of about a kilometer, instead of 5 feet, 8 inches. Even with these dimensions we remain, suspended in extragalactic space, the same insignificant beings we were before, and the scale of our dimensional relation to the rest of the universe will not thereby undergo any perceptible alterations. But with these new eyes we have much truer vision. We immediately see, in all their splendor, the nearer galaxies; a great number of galaxies, even distant ones, reveal to us their spiral arms strewn with nebulae and stars; we see galaxies linked by bridges of stars, intergalactic clusters not that far away, dwarf galaxies, compact galaxies; closer by we descry rich clusters of galaxies in which we recognize the individual components, differing among themselves in appearance and size, and other, extremely distant ones in which the individual galaxies are reduced to scarcely visible points while the entire cluster appears faint and sparse, like the scantier star clusters as seen from the earth. And this space, which seemed to us empty and flat, now appears as it really is: full and deep. For our magnified eyes see the vault of the sky swarming with galaxies. A few large ones give us the correct impression that they are nearer and, along with the others (the more numerous the smaller they get), indicate the various planes of perspective of a limitless space. It is as if we were standing at the edge of a well: a comparison between the closer stones and bricks and the more distant ones, with the smallness of the mirror of water at the bottom, give us an immediate sense of its depth. The galaxies are the stones and bricks, but in this case we do not see the bottom, and the well does not lie in a single direction: it is everywhere. With our own eyes we saw the immensity of the void; with our present eyes we perceive an unfathomable but extremely crowded abyss. This abyss is the cosmos.

We wish to see more. We want to follow this vision in time. Let us imagine, then, that we can alter the frequency of our sensations so as to see every event taking place much more rapidly. We begin to perceive that the entire world of the galaxies is in motion. We see spiral galaxies whirling around like Catharine wheels; supernovae that quickly ignite and are extinguished; whole galaxies exploding; galaxies that approach one another, tearing gas and stars apart in huge tides, then separate again, scattering and dissolving a part of themselves in space; nuclei of galaxies with fluctuating brightness; quasars pulsating, flashing, exploding, expanding until they become galaxies; galaxy clusters moving and expanding.

We had seen the cosmos as an immense entity; now we see it alive. And all this is nothing compared to the most impressive motion of all. If we watch the closer galaxies carefully, we see that they are all slowly growing smaller because they are all moving away from us. Let us watch the more distant ones: they are doing the same thing, only more quickly. Let us check those much much farther away: we see them escaping at even higher, unimaginable, dizzying velocities. Thus in the way more powerful eyes showed us a bottomless abyss, the acceleration of events is now showing us an increase in velocity, the limits of which are even more inconceivable and dreadful than those of space. And if, a moment ago, as our eyes perceived a limitless world, we asked ourselves, "But can it be that there is no end to all this? And if it does end, what lies beyond?" so now, watching the flight of the galaxies, we ask ourselves in amazement, "But where will they end up?"

These are extremely difficult but basic questions, and we are certainly not the first to contemplate them. Nowhere in the world are there people who have not asked themselves these questions, at least at some point in their lives, or who do not sense them lying, at least in the dormant state, in the innermost recesses of their minds. There has never been an epoch or a people that did not seek a solution; such solutions have often been found in philosophical speculation or religious revelation. And even today the questions of the extent of the universe, of its evolution and origin, are undoubtedly the most difficult and fascinating in all astronomy; indeed, they constitute, and will always constitute, its very essence; and because of their complexity, they form what is a science in its own right: cosmology.

We also shall tackle them and, guided by the more inspired interpretations of recent discoveries, we shall try to catch a glimpse of the solutions that modern science has to offer.

Let us start with the question of the extent of the universe. The first to seek a modern solution was Newton, who, allowing for the universal law of gravitation, which he himself had discovered, came to the conclusion that the universe cannot be finite; otherwise it would all "fall" rapidly toward its center. In the last century W. Olbers showed that the universe cannot be infinite, either, and contain everywhere stars and galaxies; otherwise the nighttime sky would be extremely bright, brighter even than the daytime sky, in which the sun itself would be indistinguishable from the background. Solutions were proposed for explaining these paradoxical and contradictory results, but we shall not discuss them, preferring to dwell instead on the essential point that they are contradictory.

When, in 1917, Einstein tried to construct his first cosmological model, based on the general theory of relativity and starting with the hypothesis of a static universe (the flight of the galaxies had not yet been discovered), he arrived at two important conclusions: (1) the universe cannot be infinite; (2) the universe cannot be finite and surrounded by an infinite vacuum. The contradiction was thus corroborated and confirmed.

The problem seemed insurmountable. But there was another fundamental point to consider.

Up to this moment we have always tacitly assumed that in the space in which we move the geometry that we learned in grammar school holds everywhere—that good old geometry in which we continue to believe because it has never let us down, and which, because it was systematized by Euclid over two thousand years ago, is called Euclidean. This geometry is based on a certain number of postulates, suggested by the evidence of the facts, taken together with certain principles of logic. For example, one of Euclid's postulates states that, given a straight line and a point outside it, one, and only one, straight line parallel to the first can be drawn through the point. This is evident as long as we are sitting a desk, drawing figures, and we think that their validity extends somewhat farther, but we cannot prove this; thus the supposition that it is valid everywhere is only an abstraction on our part, just like the straight lines and the point.

Einstein supposed, then, that space is not Euclidean but finite and curved, in such a way as to be closed like a sphere. The fact that it is finite does not imply that it is also bounded. In order to better understand the situation in which we find ourselves, let us try to imagine a bizarre case. Let us suppose that strange two-dimensional beings exist on a plane. Their universe is composed solely of whatever belongs to that plane; indeed, it is that very plane. Undoubtedly, these beings also ask themselves the same questions that we posed just now, about the limits of the universe. Since, in traversing space in all directions (while always remaining on the plane, of course), they will never encounter a boundary or a break, neither will they succeed in finding conceptual limits to this activity; they will state, correctly, that their universe (that is, the plane) is infinite and unbounded.

Let us suppose that we now transport these poor two-dimensional guinea pigs to the surface of a sphere, without informing them, of course, that the "plane" on which they now find themselves is curved. Repeating the same experiment, they will traverse the entire surface far and wide without ever finding a boundary, and they will be correct in stating, once again, that they are in an unbounded space, but they would commit a grave error if they held that that space is also infinite; and we, who are comfortably watching that curved surface from outside, in a third dimension, can perceive their error without the slightest conceptual effort.

Our own case is perhaps analogous to this, and what seems to us infinite in our three-dimensional space is instead only boundless. We concluded our discussion of the continuation of our voyage in space with the thought that to find out where a road leads, the best thing to do is to follow it. Our idea was not a good one, for we did not suspect that it might have also led us back to our point of departure, or onto other roads; each road in turn would have led us to still other roads, and we could have followed endless itineraries on a still finite world.

Today the great majority of scientists believe that the space in which we live is actually curved.

What is in dispute is the type of curvature. We shall not go into the question, but it is important to point out that cosmologists foresee observations that might reveal what kind of universe we live in. Unfortunately, our present methods of observation have only just now reached the lower limit at which the observational differences between one type of universe and another start to become obvious.

Instead let us turn to another question regarding the flight of the galaxies. We have often wondered how all the galaxies recede from us at velocities that increase in proportion to the distance, what drives them, and where they will end up in this furious race. The existence of curved space now suggests an interpretation, which, moreover, may also hold in the case of space that is not curved. It is not that the galaxies are fleeing in space; rather, the space in which they are located is expanding. The galaxies are all fixed, but the universe is expanding, dragging them along with it. This is why we have the impression of being at the center of the expansion and why we see all the galaxies receding from us at velocities proportional to their distance.

To clarify this concept further, let us turn again to the example of the sphere. Let us imagine that we take a balloon, mark a number of equidistant points on it, and blow it up, steadily expanding it (Fig. 10.7). The points will recede from each other at the same rate, for the expansion of the balloon is uniform. Let us imagine now that on one of these points there is an insect that does not know it is on an expanding balloon but sees the points that are around it. Well, that insect will think that its position remains fixed while all the other points move away from it, and it will have the impression that the more distant points are receding more rapidly, which is exactly what happens with the other galaxies in relation to our own.

This interpretation implies a consequence: if it is true that the universe is continually expanding, this means that in the past it must have been much less extensive than it is now. Indeed, there must have been a time when the matter now scattered in the innumerable galaxies, stars, nebulae, and planets was all concentrated in a relatively restricted volume.

But at this point some modern cosmologists stop: "Wait a minute," they say, "what makes you so sure that all the matter you see has always existed, or, if it has not existed for all eternity, that it was all created together?"

CONTINUOUS CREATION

It is now clear that, at the point we have reached, we can no longer go ahead relying solely on the principles and laws of physics introduced while we were experimenting on Earth or in its vicinity.

At the beginning of this century atomic physics, which was just then developing, underwent its severest crisis, which was resolved only when scientists realized that, when descending into the world of the extremely small, one cannot transpose, unchanged, the laws found in our own world. Thus it was that the indeterminacy principle replaced the principle of causality, and the concept of certainty became only a particular instance of the concept of probability, which remained the sole ruler of the world of the atom.

Similarly, in developing theories that embrace the entire universe, while retaining and applying all the laws we have discovered when we refer to the parts that compose it (be they atoms, planets, stars or galaxies), we must introduce new hypotheses or define new principles that refer to the whole. Obviously, even in this case the hypotheses and principles will be reduced to a minimum and will result directly from the interpretation of experience, as a pure abstraction of what we observe or think we observe. It is to be expected that if the hypotheses or principles we start from are different, the ensuing cosmological theories may be different as well. From now on, therefore, we shall no longer have before us results that are certain (at least conceptually); rather, we shall have various results that we shall accept or not accept, depending on our confidence in the premises or the credibility of the results.

This having been said, let us pause for a reflection that is simple but of vast import. As everyone knows, the ancients believed that the earth was the center of the universe and that the sun, the moon, the planets, and even all the stars (which were pictured as being attached to a sphere) formed a ring moving around it. Four hundred years ago Copernicus wrested the earth from this privileged position, which came to be occupied by the sun instead. Later it was realized that even the sun, one star among many, had no other privilege than that of being the center of a system of bodies, a privilege it shared, however, with planets surrounded by satellites and perhaps with a great number of other stars, centers of as many planetary systems. Nevertheless, at the beginning of the last century the research of W. Herschel led to the belief that our solar system was located in a central position with respect to the enormous stellar system of which it was a part—the Galaxy. I have already explained this error and how it is possible to determine our real position in the Galaxy. But the discovery of the flight of the galaxies, all moving away from us, placed before us the possibility that we inhabit a special galaxy, at the center of the universe. We just learned that this is only an illusion due to the fact that all of space is expanding, and that the same effect is apparent to anyone in any other galaxy.

Thus every time we thought that we were in a special place in the universe, we were mistaken, because we later learned that the earth is not the only planet, the sun not the only star, the Galaxy not the only galaxy, and that neither the earth, the sun, nor the Galaxy occupies a privileged position.

Furthermore, today we know that any body, for example, a lump of iron, is composed of an enormous number of atoms and molecules, but if we break it up into a number of pieces, each piece will be, in its essentials, like any other. The same holds true for the universe, provided that we consider sufficiently large volumes. In other words, if we take small parts of space, we may find in each of them a certain number of planets, or a few stars, or even a few galaxies. But if we consider a volume greater than that containing one galaxy or a galaxy cluster, we shall find galaxies of all types; more important, if we consider equal volumes in different zones of the universe, even at very great distances from one another, we will find the same average density of matter, and matter will have the same properties. This concept expresses the so-called cosmological principle, which, stated in its most condensed, precise form, says: *On a large scale the structure and the properties of the universe are everywhere the same.*

It is important to stress that, although it is suggested by experience, this principle also arises from a purely rational exigency. Scientists need it, for they need to believe in some foundation on which to base their entire edifice, just as they need the principle of the conservation of something (whether the motion or rest of a body, or matter, or energy). This cornerstone, which supports the entire construction, cannot be provided by experience, which can only direct reason on the road that leads it to formulate the principle. The principle itself remains, on the other hand, an abstraction, of which reason alone is the sole ruler and judge. Therefore, a principle that appears reasonable to some scientists may not be so for others, and they can refute it or replace it with another that satisfies their reason.

This has even happened in the case of the cosmological principle, for which, some twenty-five years ago, H. Bondi and T. Gold proposed to substitute the so-called perfect cosmological principle, which states: *On a large scale the structure and properties of the universe are the same everywhere and at all times.*

This principle has its origins in observations which imply the uniformity of the universe on a large scale. Other considerations have established the impossibility of a static universe, whether infinite or finite. Thus for the concept of a static universe the English cosmological school substitutes that of a *steady-state* universe in which (we are still considering large volumes) there are no changes in its structure or properties not only in space, but also in time. In short, the universe is like a stretch of river in which, though the same water never passes through it twice, there is always the same quantity of liquid, with the same properties and the same average number of waves, which have the same shapes, etc. The flow of the water in the river corresponds, in this case, to the expansion of the universe, which is also accepted in the steady-state theory of the universe.

In this case, however, the dilatation of space entails two important consequences.

Let us start with the first. As we have already seen, observations of the most distant galaxies show that their velocities are very high, approaching that of light. Given that their velocities increase with their distances, one can imagine galaxies at the limit implied by the velocity of light and beyond. We shall never be able to observe such galaxies, for the light that would inform us of their existence can never reach us. In other words, these galaxies leave our universe and cease to exist.

On the other hand (and this is the second consequence), if we accept the perfect cosmological principle, it states that the properties of the universe are the same not only everywhere but also at all times. Thus we consider a certain volume of this expanding universe: if at a given instant it contains a certain number of galaxies (for example, fifty), after some time these fifty galaxies will have moved away from each other on account of the expansion and in the same volume there will remain a lesser number—forty, say. This means that the density in our sample of space has diminished, and this contradicts the perfect cosmological principle, according to which the properties of the universe do not change with time.

Bondi, Gold, and Hoyle assert that for the density to remain unaltered and for the perfect cosmological principle to be maintained, there must be a continuous creation of matter in space (Fig. 10.8). The quantity that must appear in the universe for equilibrium to be maintained is one hydrogen atom per cubic centimeter every thousand million million years. That may seem a small quantity, but let us not forget that a cubic centimeter is also very small, especially when compared with the volume of the universe. Those who support this theory do not exclude the possibility that matter may be created in large quantities and in relatively restricted zones—for example, in the quasars, which could be particular points where matter enters the universe.

Here then, summed up, is the steady-state theory of the universe, based on the perfect cosmological principle: the universe has always existed and will always exist; at each instant a certain quantity of matter or energy "leaves" the observable universe while an equal quantity, being created continuously, "enters." Of course, it is only a theory, which, starting from certain hypotheses, explains some facts, in particular, the perfect cosmological principle. We shall now see what results we can obtain when we start, instead, from the cosmological principle in its original form, which was stated first.

THE GREAT EXPLOSION

Let us suppose that the perfect cosmological principle is not true and that the universe is the same everywhere, but not at all times. This means that it has undergone and is undergoing an evolution in time. But in what time? Perhaps in that measured by our watches? Certainly not, for we have seen that such time is valid only for us and that we need only go to a star in Orion to think ourselves contemporaries of Augustus, while for a being now living on a planet in Andromeda we have a long way to go before being born.

Let us try, then, to introduce a new concept of time that can be adapted to the whole universe and that we shall therefore call cosmic time. The only way to define it is by means of the evolution of the universe. The cosmological principle states that the universe is the same at all points but not necessarily at all times. If, then, we take two separate measurements of a certain physical quantity—for example, the mean density—and we obtain two different results, we shall say that between the two measurements a certain time has passed that will be measured by the variation in density. This cosmic time will turn out to be connected to the universe as a whole, and its flow will indicate the various epochs of cosmic evolution.

On the other hand, we have already observed that the universe is expanding, and if we do not accept the steady-state theory of the universe, tied to the perfect cosmological principle, we must conclude that, going

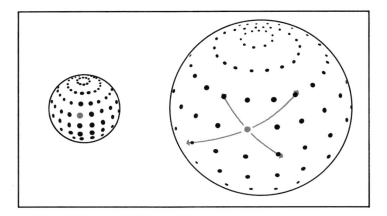

Fig. 10.7 Representation of the expansion of the universe by means of an expandable balloon on whose surface are zones of inelastic matter (represented by circular points). The gray point is the Galaxy.

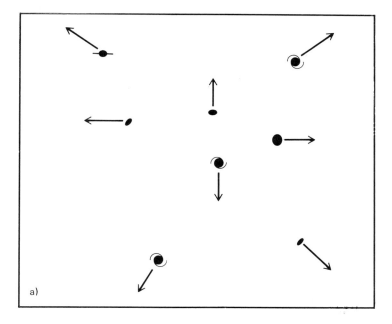

a)

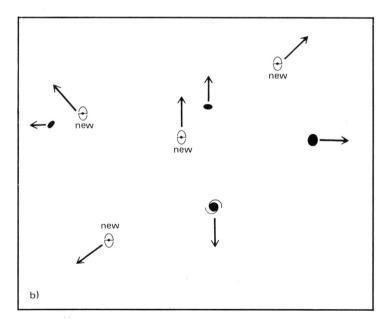

Fig. 10.8   The consequences of the perfect cosmological principle and continuous creation: if the universe is expanding and the galaxies are moving away from each other (a), the continuous creation of new matter and new galaxies is necessary to keep the density constant (b).

backward in time, there must have been a time when all matter was concentrated in a relatively restricted space. It is calculated that this must have occurred 15 billion years ago.[3]

At that time all the matter and energy in the universe must have been concentrated, at the enormous temperature of 100 billion degrees, in a sphere that astrophysicists have named the fireball. This sphere had a radius scarcely one ten-billionth that of the present universe and was composed of protons and neutrons, radiation of extremely high density, and a gas of photons, electrons, and neutrinos. Although it was in thermodynamic equilibrium, this fiery mixture could not remain static for an indefinite length of time (as the astrophysicists have shown); thus, at a certain moment, it must have begun to expand. It is at this point, with a change taking place, that we can begin to speak of time.

It is the origin of time. It is the origin of our universe. It is the first day of creation, when the amorphous, primitive chaos begins to move toward becoming something. It is the first step taken by the cosmos (which is still a crucible of matter and energy) toward more evolved forms, toward the formation of the elements, the stars, the galaxies, and farther still, up to the animated being, the thinking being, the intelligent being who will be capable of comprehending the long road from matter to himself and of following it—transcending that same matter—toward something that goes beyond his own body, beyond intelligence, something that is perhaps the origin of everything and toward which everything certainly tends: the spirit.

The magic moment of the origin! It is calculated that in only one second the temperature of the fireball would have fallen from 100 to 10 billion degrees, while the radius increased a good ten times. After scarcely a hundred seconds the radius would have increased a hundred times. This is not just an expansion, but a true explosion of the universe—an inconceivable explosion. From it the universe was born, and its dizzying course is still continuing today; we do not know when, or if, it will ever stop.

In the first few minutes, through a marked synthesizing activity of the

[3] According to results published by A. Sandage and G. A. Tamman in 1975, this event would have occurred 17,700 million years ago, if there were no deceleration in the universe's velocity of expansion; in practice there seems to be a very small deceleration. The same authors, assuming a value of 0.1, find the age of the universe to be 15 billion years.

atomic nuclei, the light elements are born and a mixture is formed, composed essentially of hydrogen (around 75%) and helium (about 25%), and containing traces of other elements such as lithium and beryllium. Meanwhile, the temperature continues to decrease: after ten thousand years it has reached 10,000 degrees Kelvin, and after a million years it is barely 600 degrees. It is at this point that, from the primordial mixture of hydrogen and helium, the first condensations begin to form, from which will later be born the galaxies, populated by stars, in whose interiors (as we have seen) are formed, from hydrogen, the heavy elements, up to iron, lead, gold.

It is a stupendous cosmic vision so fascinating, especially from an aesthetic point of view, as to seem more the work of an artist than a description of reality. The doubt arises that physicists have allowed themselves to be ruled by the poet concealed in their subconscious and have chosen the numbers so as to give life to their vision, the way musicians choose notes and painters colors.

Therefore, let us request something more—a concrete, experimental proof of what happened. When we are told that the sea once covered a certain part of the earth, we can believe it, but if we find traces of fossil fish we are even more convinced. We can try to find something of the kind as proof of the evolution of the universe as well.

More than twenty-five years ago G. Gamow had already posited that a fossil of this type must exist. We have seen that, from the moment of the great explosion, the temperature of the fireball diminished rapidly. This was natural, since the volume of the universe is steadily increasing and the thermal energy, spread throughout an increasingly large volume, must have been continuously diluted—uniformly of course—throughout the universe. But "diluted" does not mean destroyed. Today there should be a general thermal radiation scattered through the universe, corresponding to a relatively low temperature—the last remnant of that extremely high temperature present in the fireball. Well, in 1965 two engineers at the Bell Telephone Laboratories accidentally discovered this background radiation. Subsequent measurements have shown that it corresponds to a temperature of about 3 degrees Kelvin (more precisely, 2.7° K) and have confirmed that it does not come from any particular direction but reaches us in equal amounts from all directions in space, in such a way as to suggest, reasonably, that it fills the entire universe in a uniform manner.

We have found a fossil of the primitive universe, but we can do much more. When, during our journey from the moon to the galaxies, we decided to stop and to proceed farther only by looking, I assured those who had accompanied us so far that they would be rewarded by a sight much more extraordinary than the one offered by the explosion of M 82 or by an excursion to the most peculiar of galaxies. Now the time has come to fulfill that promise. When, from the earth, we observe the most distant galaxies, we reach distances of hundreds of millions of light-years; observing the quasars, we venture out several billions of light years. That is, we see galaxies and quasars not as they are now, but as they were billions of years ago. Our journey in space is turning into a journey in time, to a time that is extremely remote, for OQ 172, the most distant quasar observed thus far, is about fifteen billion light-years away. But fifteen billion years ago the universe was just being born; hence, if the "Big Bang" theory is correct, we may have a proof that is even more tangible than the three-degree radiation, as a relic of the explosion: we can observe the birth of the universe.

To reach this point, we can use the counts of extragalactic objects as a guide. When we advance to very great distances, we shall find that the number of galaxies diminishes, whereas that of the quasi-stellar sources steadily increases. From counts of radio sources we find that, at still greater distances, even these diminish. These variations in the numbers occur, not in space, but in time; as we move farther and farther away from our galaxy, we see the universe as it was a billion, five billion, ten billion years ago, and during this journey in time we reach back first to the epoch of the galaxies, then to that of the quasars, approaching that remote initial instant in which not even the quasars existed, nor even the mixture of hydrogen and helium: the fireball, the primeval chaos, the origin of everything.

Perhaps we have already reached the farthest limit of the epoch in which the quasars were formed, but let us not forget that in previous epochs evolution was much more rapid and that the universe was considerably different from the present one, especially in the first few million years of its life.

Modern instruments have already brought us very close to the moment of origin, to a cosmic time on the order of 400 million years. The coming generation of more powerful radio telescopes will perhaps enable us to come as close as only 10 million years from the initial instant. Another step and we shall observe the fireball that we have already reached and observed by detecting the radiation of 3 degrees.

And at this point we can truly say that we have arrived at the limits of time and space. We shall no longer be able to wonder what lies before or beyond, for there is no longer any before, and beyond does not exist.

CONCLUSION

Our great journey is at an end. Rapidly, as if waking from a dream, let us return to the body from which we started: the moon, or better yet, the earth, which we find much more comfortable. And now, still under the impression of a glimpse (albeit brief and partial) of extraterrestrial reality, let us ask ourselves, "What does the conquest of the moon or of a planet of the solar system represent for man in the context of an assault on the universe?" Nothing—merely the conquest of another grain of dust, at a distance corresponding to a second or a few seconds, compared with billions and billions of other grains scattered everywhere, lost in abysses millions and billions of light-years deep. It is clear that, relative to the vast spaces with their myriad worlds (which through significant glimpses we have divined rather than seen), even to reach the nearer stars and to explore their planets would still be nothing.

One could hope, of course, that the nearer stars would be merely an intermediate stage and that then, by moving from one planetary system to the next or by means of huge, nonstop leaps in space, man could reach distances from earth so great that they would no longer be insignificant even on the cosmic scale.

That is an illusion. Modern physics states that no body, no spacecraft, can exceed the velocity of light. Thus in order to reach heavenly bodies tens or hundreds of light-years away (assuming that one has constructed a vehicle for such an enterprise), it would be necessary to make extremely long voyages with a crew that would have to renew itself for many generations. This crew would have a hard time maintaining—or reestablishing—a link with people back on Earth. We must bear in mind that the men and women of the later generations, born and brought up on the spaceship, would consider the ship, and it alone, to be their world. Their culture would be that of Earth at the time their spaceship left it, and they would have acquired a store of knowledge, customs, and sensations—accumulated in space from birth or transmitted by their forebears—all completely new and inconceivable to an earth-dweller. As for the terrestrial culture itself, stowed away in their spaceship in the form of films and magnetic tape-recording, what meaning could it have for someone who has never lived on Earth? What would a view of the Matterhorn or of a fiord signify to someone who has never seen them, except as images a few meters or so high? What poetry could the inhabitants of the spaceship find in the finale of the first act of Shakespeare's *Romeo and Juliet,* when they have never seen a dawn and when a lark is merely an insignificant animal they scarcely recognize from microfilms of terrestrial zoology? Instead, superimposed over the steadily fading memories of their original world would be new experiences and traditions, derived from the distances covered, from worlds newly discovered—the results of tragedies suffered or victories gained by a new microcosm of humanity.

In the meantime earth-dwellers will have also made considerable progress but, given the enormous difference in the environment, in a completely different way.

Let us suppose, however, that such a voyage is possible and that the ship has traversed enormous distances, reaching, for example, a planet in the Small Magellanic Cloud. If the ship were traveling at a velocity close to that of light, the journey there and back would take four hundred thousand years. Assuming that all technical difficulties had been overcome, in terms of the spaceship reaching its destination and the survival of the crew (or rather its survival for five thousand generations or more), would they be able to find their way back to the solar system and the earth? And even if they could, what would be the use, since, after about half a million years, the inhabitants of the spaceship would find a world different enough from the one their ancestors left behind to be practically another world? Finally, who on Earth would be awaiting these descendants, considering that we, their common ancestors, would be as far removed from them as the later australopitheci are from us? And might not these descendants be so changed as to resemble extraterrestrial beings to their Earth-based cousins?

This voyage would be only one step in the cosmos, a very small one and in only one direction, since beyond the Magellanic Clouds, beyond the Galaxy, at enormously greater distances, this inconceivable adventure could be repeated thousands, millions of times.

Thus it is childish to view the moon landing as the first step in the conquest of the cosmos. The race to rule the cosmos is senseless, and the moon cannot be considered the first step of a ladder if the ladder vanishes after a few steps. For humanity, the moon is part of a less impractical and much more important dream. Man's arrival on the moon may mark a new era in the history of humanity, in two different but interconnected ways: it makes us citizens of space and opens the door toward a new knowledge of the world.

With the first spatial enterprises man no longer belongs to the earth; he is already expanding and living in almost a new dimension. Until the Renaissance, except for a few isolated intuitions, Man had lived in a dull world. Galileo, Newton, and Herschel turned him into the observer of a universe of dizzying and hitherto unsuspected depths. Now he has begun to shift the point from which he observes that same universe. It is a small shift: from the moon the starry sky still appears the same. But already the eclipses are different, above and below have a different meaning, and the earth is no longer the base for observations but an object to be observed. In this new dimension we are not only looking, we are moving. The first astronauts reached the moon, the next will land on other planets or satellites of the solar system, and those of a more distant future perhaps on some planet of one of the closer stars. They will set up bases in space and perhaps found colonies, and people will think of them not as "up there," but simply as "farther out."

The conquest of this new dimension will not occur—is not occurring— solely through material expansion in space but also through the consciousness of understanding, because from the moment that we become aware of it, all of us (even those who stay on Earth) cease to be inhabitants of the old, dull world and become space travelers. Those on Earth consider themselves inhabitants of the earliest, largest base, (which nonetheless, like all the others, is suspended in the void and moves through it at a high speed), the base on which everything depends, the base they not only inhabit but for which they are the responsible guardians.

Such awareness on the part of all humanity, which the single exploit of *Apollo 11* imparted almost instantaneously to all those who take an active interest in the life of this world, is the starting point that will enable Man to commence his true, conscious expansion in space. It is an expansion full of unforeseeable consequences, an expansion inspired and stimulated by two great passions that have spurred man on ever since he appeared on Earth: the spirit of adventure and the thirst for knowledge.

This is a terrible thirst which constantly torments man and is forever being slaked. Let us not forget that the universe that crushed us just now with its immensity and with the majesty of its phenomena has been slowly discovered, measured, and analyzed by man, and it is only recently that our imagination has grasped it, with a supreme effort.

Moreover, nearly everything that we have seen was discovered from Earth. We have already begun the geological study of our satellite and of the nearer planets, particularly Mars. Before long we shall succeed in installing astronomical observatories on the moon, where, by virtue of the absence of an atmosphere, heavenly bodies can be studied in a way that would never have been possible from the surface of the earth. The absence of scintillation, the greater duration of observations (never interrupted by clouds), the possibility of detecting radiation over the entire spectrum—these factors will bring about progress in our knowledge of the universe as great as that instigated three centuries ago by the introduction of the telescope. Our increasingly thorough exploration of the cosmos will reveal surprising facts and physical phenomena. Moreover, once we can travel even to the closer heavenly bodies (other than the moon, Mars, and Venus), we shall be able to determine whether there are living beings there, of some kind, which have, like those beings that populate the earth, the fundamental properties of growth, reproduction, sensation, and, perhaps, thought. How important such a discovery would be—to find a proof that all that we have seen is not merely inert matter, devoid of self-awareness, but the setting appropriate to the development of a cosmic life.

Thus we can now better appreciate where that first step on the moon led us—a step watched by a large portion of the world on that memorable night in July 1969, yet without the full realization of what it meant. It led us, above all, toward a new door, which we have slowly begun to open, knowing that beyond it lies the coveted prize: knowledge, the noblest form of conquest, which has no need of a crushing heel but requires only an attitude that incorporates understanding and appreciation.

APPENDIXES

A THE DETERMINATION OF THE DISTANCES OF THE STARS

The principal objection that the Ptolemaists made to the Copernican system was that if the earth were really moving round the sun, the motion should be reflected in a similar motion of the fixed stars.

The objection was legitimate, but equally legitimate, albeit unaccepted, was the reply Copernicus and Galileo offered: that there really was a displacement of the stars but that it was not perceptible with the instruments then available, because of its extreme minuteness. With the adoption of the Copernican system this explanation was accepted, and from the impossibility of measuring the displacements of the stars it was deduced that their distances from the earth must be enormous.

Finally, in 1838, Bessel succeeded in measuring for the first time the minute displacement of the star 61 Cygni, from which he deduced its distance, thus introducing a new era in the history of astronomy.

The procedure, known as the trigonometric method, is as follows (Fig. A.1): Let us suppose that we are looking at a star at a given time (in January, say) when the Earth is at A; it will be seen at a point A' on the sky. Let us return to observe the same star after six months; since the earth will have shifted to B, the star will appear at point B', because of the change in the position from which it is observed. The idea, then, is to determine the angle ASB made by the lines joining the earth and star, or rather the angle π, which the semi-major axis of the earth's orbit subtends at the star. This angle is called the annual parallax, and from it, one can derive the distance by using a simple formula of plane trigonometry.

One has in fact, $\tan \pi = a/r$, or $r = a/(\tan\pi)$, and since a, the semi-major axis of the earth's orbit, also called the astronomical unit, is known, the distance r of the star is directly obtainable.

Thus the method is simple in principle. The chief difficulty lies in the actual measurement of the angle π. To give an idea of how difficult this is, let us state at once that no star is near enough to have a parallax of 1″ and that, on the other hand, the angle 1″ is that subtended by 1 meter observed at a distance of 206 kilometers.

Since the distances of the stars are so enormous, it is necessary to introduce a new unit of measurement: the light-year. It corresponds to the distance that light travels in a year and is given, in kilometers, by the number of seconds in a year multiplied by 300,000 (the number of kilometers that light travels in a second). Thus 1 light-year = 9,460,500,000,000 kilometers.

No star is as near as a light-year from the earth. The nearest is Alpha Centauri, 4.3 light-years away.

For practical reasons astronomers nowadays use as the measuring unit the distance at which the semi-major axis of the earth's orbit subtends an angle of 1″. This distance, called a parsec (an abbreviation of *parallax* and *sec*ond), is equal to 3.26 light-years. For great distances, the kiloparsec (1,000 parsecs) and megaparsec (1 million parsecs) are used. In this book I consistently use the light-year, since it conveys the idea of distance more directly and also because it helps to give an idea of the temporal consequences of space travel.

Unfortunately, the trigonometric method, which is so rigorous and precise, is practically unusable for distances greater than 300 light-years, because of the extreme smallness of the angle to be measured. Let us examine this difficulty. From figure A.1 we see that to obtain π we must measure the angle *ASB* or the angles *FAS* and *FBS,* from which one can obtain π directly, the sum of the three angles being equal to 180°. Since π is very small ($< 1″$), the two angles *FAS* and *FBS* are very nearly 90°; that is, the lines *AS* and *BS* are almost parallel.

Astronomers succeed in measuring such small angles up to distances a hundred times greater than a parsec. Beyond this limit, they are unable, because of instrumental errors, to distinguish the point in which the two lines meet for they now appear to be parallel.

It is necessary, then, to resort to other methods. The geometrical methods available include that of secular parallaxes (based on the displacement of the whole solar system in space), that of group parallaxes (applicable to clusters), and that of dynamical parallaxes (applicable to double stars). Apart from these, the others are all based on the determination of absolute magnitude.

We have seen that the absolute magnitude of a star (expressing its intrinsic brightness) is the apparent magnitude that the star would have if it were placed at a distance of 32.6 light-years, that is, at exactly 10 parsecs. Now the apparent magnitude, the absolute magnitude, and the distance are connected by the equation

$M = m + 5 - 5 \log r$ (where *r* is the distance in parsecs)
or $M = m + 5 + 5 \log \pi$ (where π is the annual parallax in seconds of arc).

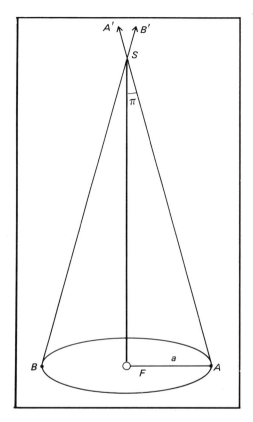

Fig. A.1   Geometrical determination of the distance of a star S from the earth by the parallax method.

By using one of these formulas, one can deduce the distance of a star if its absolute magnitude is known. The procedure is as follows: Let us suppose that we can determine, by a trigonometrical method, the distance of all the stars of a certain spectral type (for example, nongiant G stars like the sun) that are near us. From one of the two formulas we at once obtain the absolute magnitudes, which will be more or less equal. We can assign the mean of these values to all nongiant G stars that we can recognize as such from their spectra. We then obtain the distances, knowing only, in addition to the spectrum, the apparent magnitude.

Similarly, it is possible to determine the absolute magnitudes of stars of other spectral types. In order to obtain the parallax in seconds or the distance in parsecs, it is enough, then, to be able to determine the absolute magnitude of a star from its spectrum. This method (called the method of spectroscopic parallaxes) is applicable to all normal stars, as long as they are bright enough so that their spectra can be determined.

The intrinsic brightness of a star (that is, its absolute magnitude) can also be determined by other methods, always, however, starting from the comparison with stars whose distances have been obtained by geometric methods and which serve for the calibration.

Naturally, the greater the distance, the lower the precision with which we are able to measure it, but what counts is the precentage error. It would not be very serious if we gave the distance from Rome to Milan with an error of 5 kilometers; the (percentage) error that we make in estimating the distances of the stars by these methods is generally no greater than this.

Thus we can pursue our course among the stars with confidence. The problem is more difficult when we wish to measure the greatest distances that can be reached in the universe. And it is indeed to the solution of this problem that some of the leading modern astronomers are devoting so much energy and time with the world's largest telescopes.

As we have seen, the first variable star (not counting the novae) was discovered by the Protestant minister David Fabricius in 1596 and was considered so exceptional as to merit the title of Mira ("Wonderful"). Seventy years later Gemignano Montanari of Bologna discovered the second variable, Algol, and after another century, in 1784, Goodricke discovered the variability of β Lyrae and δ Cephei. From that time on, the number of these strange stars began to increase more rapidly, and by 1786 E. Piggott published a first list of twelve. Nevertheless, progress was still very slow, so much so that Argelander's catalog, published in 1844, another half-century later, contained only six more. But twelve years later Pogson's catalog contained fifty-three, and Pickering's, which appeared in 1903, reported seventy-one. In the next dozen years the number of variables more than doubled—not even counting those which had begun to be discovered in globular clusters and in the Magellanic Clouds, numbering over five hundred.

This considerable increase was due to the new method adopted in the search. Up to the end of the last century the hunt for variable stars was conducted visually, by carefully comparing various areas of the sky observed with the naked eye or the telescope to their respective charts. The comparison was made star by star and went as far as magnitude 9, that is, the magnitude of the faintest stars contained in the Bonner Durchmusterung (BD). At the end of the nineteenth century photography came into use, permitting more exhaustive and objective comparisons. With this system the state of a certain zone of the sky at various times is recorded on photographic plates that enable one to register much fainter stars than those marked on the charts. Comparing pairs of plates is simpler, more convenient, and more dependable than comparing the sky with a chart, since it can be done in one's study, using two equally reliable documents, and can be repeated at any time; the chart, on the other hand, might contain an error and the eye might make a mistake. Moreover, the search is much more rapid, since the comparison is not an individual one between star and star, but an overall comparison of the two plates. There are various methods for making the comparison, but two are most commonly employed: the "blink" method and the "positive on negative" method.

In both cases one starts with a pair of plates taken at different times but under the same conditions (that is, with the same instrument, the same type of plate, exposure, development, sky conditions, etc.). Obviously, all the stars that have not varied will appear on both plates as points of a certain darkness and size, equal in both cases, whereas the image of a star that has varied in brightness is recorded on one plate as a well-defined point and on the other as a fainter one. There are two methods of finding such a point among the multitude of other similar points. One consists in making a positive of one of the two plates and superimposing it on the other. In this way each black point on the second plate will exactly cover the corresponding white point on the positive of the first plate, producing a nearly uniform gray background. If one carefully examines this combination, with a lens or low-power microscope, one will see, standing out from the background, the point corresponding to the star that has varied, which will be white if the variable is brighter on the positive or black if it is brighter on the negative.

The comparison of a pair of plates can also be made with an instrument known as a blink microscope, or "blink" for short. With a suitable arrangement of prisms it presents the images of the two plates, which, by means of a rotating sector that intercepts the images alternately, are observed separately and in alternation. If it happens that the pictures on the two plates coincide exactly, when the images are rapidly alternated, the eye will not perceive any difference, since the stars on both plates are identical. If, however, a star is brighter on one of the plates because of variation, it will at once leap to the eye by continually presenting small flashes. Thus, in both cases the eye no longer makes the comparison star by star, but instead traverses the entire field until the variable strikes it by spontaneously appearing.

It was obvious that with this new and extremely efficient method the number of variables discovered would increase rapidly. In fact, the first *General Catalog of Variable Stars,* published in Moscow in 1948 by B. V. Kukarkin and P. P. Parenago, listed as many as 10,862. The rate of discovery continued to grow; the third edition of the catalog, published in three volumes between 1969 and 1971, contains fully 20,300 variables, and three supplementary volumes published since bring to 25,842 the number of variables named by the end of 1975. If the increase continues at the present rate, by the end of this century the number of variables noted and cataloged will amount to fully 200,000 (Fig. B.1). This prospect is more than a little disturbing to the compilers of the catalog who, even though their number has increased from two to nine, will have more and more trouble

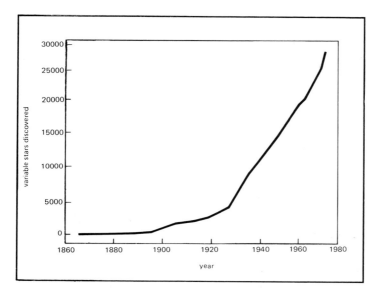

Fig. B.1   Graph illustrating the increase in the number of variable stars observed in over a century (from 1860 to the present).

including all the new variables promptly and, above all, updating the data concerning those already known, which continue to be researched by hundreds of astronomers all over the world.

As was mentioned, these variables are all galactic. Others, however, have been discovered in globular clusters and listed in scattered publications or special catalogs. The most recent catalog of this type, published by H. Sawyer in 1955, lists 1,421. Hundreds have also been found in preliminary or partial searches in other galaxies, and systematic research carried out for the Magellanic Clouds has led to the discovery of nearly 4,000 all told.

Thus if we consider all the known variables (including those outside the Galaxy) and if we recall that there also exists a catalog of "suspected variables" that contains over 10,000 more, many of which are only awaiting further study to be listed among the definite variables, it is easy to deduce that the total number of variables known so far must come to about 35,000, almost twenty times the number of stars that the eye can discern in the sky on a clear night.

We have already seen that every variable has a name, or rather a designation. The system was introduced in the last century by Argelander, who denoted with the letter R the first variable found in each constellation, the second by S, and so on, down to Z, placing after the letter the name of the constellation in the genitive (for example, T Tauri, usually abbreviated to T Tau). Argelander was too optimistic and did not foresee that the letters of the alphabet would soon be exhausted—especially if one started with R. A double-letter system was then adopted (RR, RS, RT, . . . , SS, ST, SU, etc.), which avoided using as a second letter any letter that had already served as a first letter; for instance we have the combination RS but not SR. In this way it was possible to denote in each constellation another 45 variables, which, together with the 9 designated with the letters R to Z, brought the total possible to 54. Soon this extension proved insufficient. A decision was then made to use the first letters of the alphabet, and there appeared variables designated as AA, AB, AC, . . . BB, BC, BD., etc., down to QX, QY, QZ. One could then denote 334 variables per constellation, but this limit was also exceeded, and it was finally understood that it was better to have recourse to series of whole numbers, which, being infinite, would constitute once and for all a satisfactory accounting system. Thus all the variables, starting with no. 335, were designated by a *V* followed by the appropriate number. Excluded from this classification are all those stars which, before they were found to be variable, had already received a designation (such as β Persei, α Herculis, etc.)—generally, though not always, by way of the Greek letters assigned by Bayer at the beginning of the seventeenth century.

Of course, not all constellations required the extension. There are some (such as Caelum and Leo Minor, with 9 and 16 variables, respectively) for which Argelander's nomenclature, or its first extension, suffices. Many others, however, are much more crowded. The constellation that contains the greatest number of variables is Sagittarius, with 3,891, followed by Ophiuchus (2,057), Cygnus (1,502), and Aquila (1,312). That seems natural enough, for these constellations correspond to zones crossed by the Milky Way and therefore extremely rich in stars.

By examining the *General Catalog of Variable Stars,* one can discover many other interesting things. For example, the variables clearly visible to the naked eye, being brighter than 3^m at maximum, number twenty-four; all the same, sixteen of these have such a small range of variation ($\leqslant 0^m.2$) that they cannot be perceived by the naked eye. Of the eight remaining, the most easily observable are Mira and Algol, which were in fact the first two to be discovered.

The *General Catalog* and its supplements also list unusual stars, such as 256 nameless stars handed down by ancient chronicles from 2296 B.C. to the seventeenth century A.D., presumably novae or supernovae; in the case of many of these we have extremely scanty information or even doubts concerning their very existence. There are also variable objects that are stellar in appearance but are certainly not stars, such as 127 optically variable quasars and 55 that include peculiar galaxies and nuclei. Finally, although ordinarily it only deals with stars belonging to the Galaxy, the catalog lists all the extragalactic supernovae, 491 in number, starting with one that presumably appeared in M 31 in 1664.

Apart from these pieces of information and curiosities, the main statistical interest of the catalog consists in an overview of the distribution of the variables among the various types. Thus we find that the most numerous by far are the pulsating variables, which total 17,327, of which 5,169 are long-period Mira-type variables, 5,894 Cepheids of Population II (of the RR Lyrae type), and 769 long-period Cepheids, mostly of Population I. There are 2,844 eruptive variables (novae, novalike, nebular variables, and flare stars) in all. Eclipsing stars of various types number 4,725. Some variables (55 in all) are unclassifiable; each one constitutes a unique, solitary case. Finally, 808 have not yet been studied enough to be classified.

The fact that so many variables remain to be studied should come as no surprise; indeed, one might add that many of those already studied were studied only partially, and more accurate and thorough observations, which are certainly to be desired, might hold some big surprises. Unfortunately, the number of variables is too great, and astronomers around the world are able to follow only a fraction of them. Luckily they are aided by a devoted band of amateurs who, seeking in the study of the heavens that relaxation and sense of relevance which seem, more and more, to be lacking in our daily life, make an indispensable contribution to one of the most important and interesting areas in the science of astronomy.

In discussing the stars, we have frequently encountered the names or designations by which some of them are known. The same is true in the case of other objects, such as the nebula M 42. This happens because, for both the stars and the principal celestial objects, there exist catalogs, general or special, containing their positions and principal characteristics, by means of which the objects can be described and located in the sky.

The oldest catalog that has come down to us is attributed to the Greek astronomer Hipparchus, who compiled it around 127 B.C.; it is, almost certainly, the one contained in the famous Almagest of Ptolemy. Listed in it are the positions and brightness of 1,022 stars. That is not a great number, but, after all, the human eye does not see many more! When Hipparchus' contemporaries learned of the catalog, they were greatly impressed. Indeed, the naturalist Pliny exclaimed, referring to the remarkable astronomer who compiled it, "He dared to count the stars and name them, a difficult task even for a god!" From that time on the number of catalogs and star atlases has steadily increased, reflecting, at various times, the three fundamental innovations corresponding to three major changes in the methods and techniques of observation.

The first fundamental change came when, with the introduction of the telescope, the eye could avail itself of a means that literally multiplied its power. The first catalog compiled with the help of the telescope was that of 341 southern stars drawn up by Halley on the island of St. Helena in the years 1677–78. The most extensive is the *Bonner Durchmusterung* (BD), still widely used, which contains all the stars north of $-2°$ brighter than magnitude 9.2 and many others down to around magnitude 10.5. The compilation of this work, which was later extended to declination $-23°$, was undertaken by Argelander and Schönfeld and took from 1852 to 1884 to complete. The catalog and the corresponding atlas contain a total of 457,848 stars, all observed and measured one by one.

The results obtained by visual observation were subsequently surpassed by the photographic plate. This method has, in fact, the enormous advantage of accumulating effects, so that, with a photographic camera attached to a telescope and sufficiently long exposures, it is possible to register faint objects that the human eye could not perceive even with much more powerful instruments. Taking advantage of this new technique, astronomers began, at the end of the last century, to prepare a great photographic chart of the sky. This splendid undertaking was so vast that it had to be divided among eighteen observatories around the world, all of which used instruments having the same power and covering the same field of four square degrees. A catalog was compiled along with the atlas. After some thirty years' work most of the observatories had finished collecting the necessary plates, and the catalog containing stars down to magnitude 12, which number about two million, was well advanced. To this day, however, the work of reducing the catalog has not yet been completed by all the observatories involved, and some observatories have not even published the charts. The astrographic catalog, initiated under the auspices of the worldwide brotherhood of science, became a great source of disappointment for many astronomers.

As far as organization is concerned, an undertaking of this kind could be repeated, but not surpassed. Nonetheless, the observation has been made that it was uneconomical, engaging too many instruments for too long a time.

Fortunately, in 1932 B. Schmidt, an Estonian technician living in Hamburg, constructed a reflecting telescope equipped with a glass correcting plate capable of eliminating those aberrations which considerably limit the performance of reflecting telescopes. The new instrument possessed two basic advantages: high luminosity and a wide field. Its introduction brought about the third decisive revolution in this area of research.

The largest instrument of this type, installed at Mount Palomar Observatory between 1949 and 1956, has produced the greatest photographic atlas of the sky, although unfortunately it is limited to the sky north of declination −27°.[1] This atlas shows millions of stars down to magnitude 21.1, and hundreds of thousands of other objects—nebulae, star clusters, galaxies, etc. Each field has been photographed in blue and red light, so that, from possible differences in the appearance of the objects or in the intensity of the stars, one can derive the color of every celestial body photographed.

In addition to star catalogs there are also catalogs of other interesting

---

[1]The atlas has been extended, with subsequent additions of lower quality, to declination −42°. The completion of the atlas, for all the zones south of the equator, will be carried out with the Schmidt telescope of ESO (European Southern Observatory) in Chile.

celestial objects. The first and most famous of these was compiled in the second half of the eighteenth century by the French astronomer Messier. He was an experienced comet-hunter, and in the course of his life he observed more than forty, of which he himself had discovered thirteen.

He compiled the catalog of the principal celestial objects (star clusters, nebulae, etc.), not to indicate the objects themselves but to help him in his search for comets. The idea first occurred to him in August 1758 when, while following a comet discovered the previous May by de la Nux, he ran into the Crab nebula. The momentary misunderstanding that this caused him led him to resolve to make a list of all those bodies which could be mistaken for comets and which, in the search for the latter, should be discounted.

In 1764 he began a systematic search, and in 1771 he published an initial list of 45. The definitive catalog was published in the *Connaissance des temps* for 1784, printed in 1781. It contains 103 objects, each of which is now indicated by the number with which it is listed in the catalog, preceded by an M. Some of these, such as the Pleiades, mentioned by Heslod and Homer, or M 44, discovered by Hipparchus, had been known since antiquity; others were discovered by Messier's contemporaries, such as P. Méchain, another comet-hunter, who discovered 22 objects, particularly in the constellations Virgo and Coma Berenices, and passed his observations on to Messier. In addition to collecting and checking others' observations, Messier himself discovered 42 objects.

Although his list contains 103 references, we now find some Messier objects with higher numbers, up to 110. That is because other authors have added clusters and nebulae later found in Messier's papers (letters, notes, etc.) to the original list. On the other hand, there are also some numbers in Messier's list that do not correspond to precise celestial bodies but merely to two or more stars apparently close to one another (such as M 40 and M 73), or do not correspond to anything (such as M 102, which is an erroneous repetition of M 101).

After this list of Messier's (which essentially comprises the brightest, and hence most thoroughly studied, celestial objects), other astronomers continued to discover and observe other objects, which they listed in various catalogs. They were all collected by J. L. E. Dreyer in the *New General Catalogue*, published in 1888, and in the two *Index Catalogues*, published in 1895 and 1908, respectively. These catalogs contain a total of 13,266 objects (including those in Messier's catalog), designated by their progressive number preceded by the abbreviation NGC or IC.

To sum up, then, modern astronomers have at their disposal a large atlas and catalog, the BD, extended by the *Cordoba Durchmusterung* to include the entire Southern Hemisphere; a splendid two-color photographic atlas from the North Pole to southern declination $-42°$; and a general catalog of the principal nebulae, clusters, galaxies, etc., denoted by NGC or IC.

In addition to these there exist numerous specialized catalogs, from the more general ones of double stars or variable stars (which we have already encountered) to other more restricted and specific ones dealing with fainter objects or more detailed information. There are also catalogs of objects discovered only recently by means of new techniques, such as those involving infrared stars, radio sources, or X-ray sources. There is no need to dwell on these catalogs, which are of interest only to specialists.

Instead let us note one last atlas, the *Atlas Coeli,* compiled together with the *Atlas Coeli Catalogue* by the Czechoslovak astronomer A. Becvar. Easy to consult and moderate in price, used by both professional and amateur astronomers, this combination atlas-catalog offers, in a practical and modern synthesis, a complete view of the sky and of the most striking or famous objects, including novae, double stars, radio sources, and dark nebulae.

D THE MAFFEI GALAXIES

The discovery of two new galaxies now known as Maffei 1 and Maffei 2 is a most recent example of one of those strange coincidences through which one finds one thing while seeking another.

Horace Walpole coined a word to represent the faculty for making discoveries of this kind: *serendipity*. The word derives from a Venetian tale of the Renaissance, which tells the story of three princes of Serendip (Ceylon) in search of a hundred verses containing the secret of a liquid capable of killing all sea monsters: they found only a few fragments of the magic formula, but in the course of their search they made many other unexpected discoveries, simply because they were looking for something. "Serendipity," then, is equivalent not to "chance" or "fortune"—or, rather, not *only* to these—but to "the gift of discovering, by accident and shrewdness, things that one was not seeking." Obviously this almost always happens when one is traveling a new road (or at any rate an unfrequented one), or using a new technique. That is also what happened in this case.

By 1956 I had already started to photograph celestial objects in the infrared. The technique had been around for a while but was not widely used; it is just starting to be now, following the enormous development of research in the far infrared (that is, at wavelengths from 1.2 to 3,000 μm), carried out with special instruments and refined techniques. These do not, however, detract from the importance of research in the near infrared (shorter wavelengths, bordering on visible radiation), which make use of photographic recording. In fact, the photographic infrared is often sufficient to suggest what happens in the far infrared, and, furthermore, it provides rapid results that are easily visualized by means of photographic images, rather than numbers or graphs.

One of the fields in which the study of the behavior in the infrared can prove particularly interesting is that of stars in the process of forming.

As we have already seen (in Chapter 5), stars are continually being born in special zones rich in gas and dust, and no sooner do they form than they appear gathered in groups called associations. Many of the stars in the associations show various peculiarities. In some of these zones, in particular, I had already been able to obtain interesting results by comparing the behavior in the blue with that in the infrared. I recall, among others, a small comet-shaped nebula in the rich zone of Orion, barely visible in the blue but intense in the infrared; at one end of it I found an infrared star of variable brightness.

Among the zones that appeared to me to be worthy of interest and not yet sufficiently studied was one in the constellation Cassiopeia, known as IC 1805; I had already taken some photographs of it in 1962, with the Schmidt telescope in Hamburg, one of the largest in the world. In the autumn of 1967, while preparing an observation program to be carried out with the new Schmidt telescope at the Asiago Observatory, I decided to make both blue and red observations of this field (Fig. D.1). I thus discovered the brighter of the two galaxies, a matter of days before the official inauguration of the instrument. The first pair of blue-infrared photographs were taken on the night of 29 September 1967 (Fig. D.2). I had a look at the plates the next day, just as soon as they were developed and dried. This is a normal procedure, but this time it was not just for a technical check, because there was also interest and curiosity aroused by something new, since this was the first pair of plates that I had taken of this field. The zone is extremely rich in nebulae visible only in red light, which disappear completely in the infrared (as is normal). Thus I was very surprised when I noticed on the infrared plate, rich in stars but absolutely without nebulae, a dark point larger than the points attributable to the stars. Under the microscope it appeared diffuse, and it seemed clear that I was not dealing with a star. The biggest surprise, though, was the fact that when I looked for it in the same position on the blue plate, I found nothing.

To remove all doubt, on the night of October 1 I took another pair of plates, and this time the small nebula appeared on the infrared plate again, with the same extension, the same brightness, and in the same place. Thus it was certainly a celestial object. But what was it?

I thought of the small comet-shaped nebula in Orion. Perhaps it was something analogous. But that nebula was visible (even if very faintly) in the blue. Here, besides, there was no infrared star that could illuminate it, as there was in the other. Perhaps it was a more mysterious and unusual body, in which the matter that forms the stars was going through a phase as yet unknown and unexplored. Perhaps I had found for the first time that missing link between star and prestellar matter that until now had always eluded us.

I had to obtain more information. I looked for the little nebula in the Mount

Fig. D.1   The discovery of the Maffei galaxies: enlargement of the field of IC 1805, photographed in red light with the Schmidt telescope of the Asiago Astrophysical Observatory. The two galaxies are indicated by arrows in the lower right-hand corner: *above,* Maffei 1; *below,* Maffei 2.

Fig. D.2 Enlargement of the region surrounding the galaxy Maffei 1, photographed in the blue (*left*) and in the infrared (*right*) with the same instrument. Note the absence of the object in the "blue" photograph. (*P. Maffei*)

Fig. D.3 Maffei 1, photographed with the 200-inch reflector of Mount Palomar.

Palomar photographic atlas; I wanted to see whether it had been recorded by others and if it was a radio source. I also examined the material that I had collected at Hamburg—but all without making any great progress. To learn more about it, I would have to carry out photometry in more colors, especially in the far infrared, and obtain a spectrum with a slit spectrograph: but for the first task we did not have the proper equipment in Italy, and the second would require a large telescope. Therefore, in April 1968 I decided to publish what I had discovered, hoping that others would carry on the research.

It was then, while I was preparing the photographic material to illustrate the notice, that I discovered the second object, much fainter than the first but with the same strange behavior in the blue and infrared.

I sent the notice of the discovery of the two objects to the Astronomical Society of the Pacific, which published it in the October issue of its "Publications."

Meanwhile, another possibility occurred to me: that the two objects might be of a very luminous type already known (such as a galaxy or a globular cluster) but that their light appeared to us extremely weakened and reddened because of the enormous quantity of interstellar dust, closer to us and interposed in great abundance in precisely the direction in which they appear. This point of view was also suggested to me by Gavril Grueff, a radio astronomer in Bologna whom I had told about the two objects; on 1 December 1969 he informed me that he had discovered that Object 1 was not a radio emitter but Object 2 was.

Robert Landau, a student at the University of California who had read my note and exchanged letters with me, resumed the observation of Object 1. The research then interested other astronomers, considered to be among the most experienced specialists, who tackled first Object 1, then Object 2 (Figs. D.3, D.4) with the biggest telescopes in the world, including the 200-inch reflector of Mount Palomar. Thus in early January 1971 nine astronomers, led by H. Spinrad and including Landau, published the result of their research on the first object and some preliminary notes on the second.

According to them it is a question of two new galaxies (to which they gave my name), greatly obscured by the interstellar matter of the Galaxy interposed along the line of sight from the earth, since they, like us, happen to lie near the Galaxy's equatorial plane, which is rich in dust. Fortunately,

Fig. D.4   Maffei 2, photographed with the 200-inch reflector of Mount Palomar.

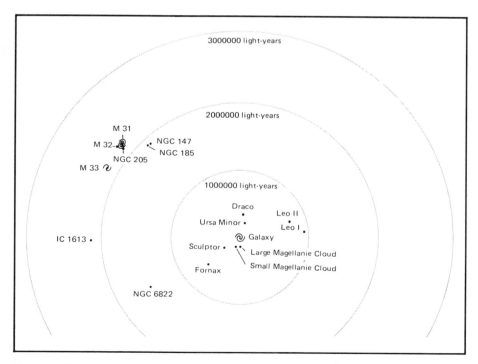

Fig. D.5   The local system of galaxies before the discovery of the two Maffei galaxies. According to the initial measurements, both the new galaxies were located not far from the group of M 31 (Andromeda galaxy), a little beyond the third circle.

they are in the opposite direction from the center of the Galaxy, that is, in the direction in which, moving out from the earth in the plane of the galactic equator, we would traverse the minimum thickness of dust before leaving the Galaxy.

The American astronomers found that the infrared emission of wavelength 2.2 μ m from the central zones of Maffei 1 is equal to that emitted by the same area of a central zone of the galaxy M 31. Then, allowing for the interstellar absorption, so that the resultant color is that of a giant elliptical galaxy and of the nucleus of M 31, they found an absorption of 5.1 magnitudes. The spectral distribution of the energy, from the blue to 3.5 μ m, agreed perfectly with the known distribution of M 31. This was the first concrete proof that object Maffei 1 is a galaxy. A second was provided by the spectra, which showed the magnesium triplet (very common in galaxies) and the sodium lines and titanium oxide bands, which have been found in the nuclei of giant elliptical or spiral galaxies.

From these and other observations it was concluded that Maffei 1 is a giant elliptical galaxy, of the E3 or E4 type, with an absolute magnitude $M_v = -20.5$, located at a distance between 0.3 and 4 million parsecs—probably 1 million parsecs, corresponding to 3.3 million light-years. It appears to be moving, with respect to the Galaxy's center of gravity, at a velocity of 165 kilometers per second, and it is believed to have a mass 200 billion times that of the sun. Thus it is a galaxy as huge as ours, albeit of a different type, and it contains at least a hundred billion stars.

Much less was known about Maffei 2 at the beginning of 1971. It seemed clear, however, that it is a spiral galaxy, and it was believed, given the small apparent distance separating it from Maffei 1, that it was physically associated with the latter, that is, that it did not appear close to Maffei 1 solely from the effect of perspective. In that case, if one adopted for both the distance of 3.3 million light-years, it followed that both belong to the so-called "local system" of galaxies, made up of sixteen galaxies arranged at various distances around our own, within a radius of a little over 3 million light-years (Fig. D.5). The importance of this discovery consisted in having augmented not the total number of galaxies in this system (which in early 1972 was further increased by van den Bergh, with the discovery of four dwarf galaxies near M 31) but the number of larger galaxies, which, with Maffei 1, passed from two (the Galaxy and M 31) to three.

Furthermore, with the addition of this galaxy of great mass, the system became dynamically unstable, a fact of great importance not only in the study of the structure of the local system but also in the study of its evolution.

Contemporaneously or subsequently, other scientists undertook the study of these objects, obtaining new results that sometimes contradicted the earlier ones. The Dutch astronomer Oort, using the new radio telescope at Westerbork, constructed a radio map of the zone at the 21-centimeter wavelength, which confirmed that Maffei 1 did not appear to emit radio waves. This result had already been obtained by others, but, thanks to the special sensitivity of his instrument, Oort was able to reach a very low level of reception. Thus he was able to ascertain that the radio emission—if indeed there is any—must be at least ten thousand times weaker than the minimum emission hitherto recorded from giant elliptical galaxies. In order to find out how exceptional this case is, a survey was started at Westerbork of all giant elliptical galaxies, measuring for each the intensity of the radio emission.

Meanwhile, important research was being carried out by other radio astronomers on the galaxy Maffei 2. L. Bottinelli and six other French radio astronomers made observations of the 21-centimeter hydrogen line and of the adjacent continuous spectrum using the Nançay radiotelescope. They thus found the velocity at which Maffei 2 is moving away from us and, more importantly, discovered that it rotates on its axis at a maximum velocity of 200 kilometers per second. Then, by various methods and starting from different hypotheses concerning the type of galaxy, they determined the distance. Assuming that it is of type Sb, as suggested by the Mount Palomar group, it would turn out to be at a distance of 2.7 megaparsecs. In that case it would be an average galaxy with a mass forty-six billion times that of the sun—that is, less than a quarter of that of Maffei 1.

Clearly, if this galaxy is really so distant, it can no longer be said to belong to the local system. According to the French astronomers the two galaxies (which they still think are associated with each other) would form part of another system, that of Ursa Major–Camelopardus to which M 81 and M 82 also belong (Fig. D.6).

Yet if we take as the common distance that found for Maffei 2, 2.7 megaparsecs, then the galaxy Maffei 1 will have a mass almost 900 billion

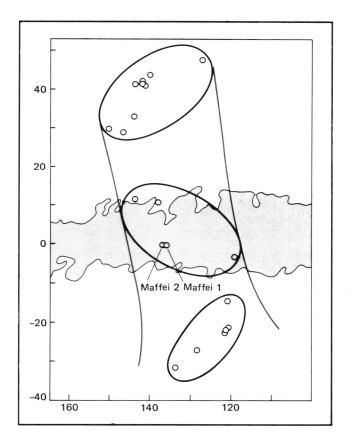

Fig. D.6   The Ursa Major–Camelopardus system of galaxies according to the observations of the French radio astronomers at Nançay. The group within the closed
curve above is that of M 81; that inside the closed curve in the middle is the Maffei
group; that in the closed curve below is that of M 31. The lighter lines indicate the
limits of the supergroup complex. The zone in gray is that of the Milky Way, where
the light absorption is very great. The abscissa is galactic longitude, the ordinate
galactic latitude.

times that of the sun. If it were shown to be a giant elliptical galaxy, it would be even harder to explain the absence of radio emission. Indeed, if the research underway at Westerbork shows that there is no giant elliptical galaxy with such weak radio emission, we would have to conclude that Maffei 1 is a dwarf elliptical galaxy and much nearer than 1 megaparsec. Obviously, in that case, the two galaxies would be separated by a much greater distance than that separating the nearer one from us, and we could no longer speak of an association between the two.

It may be that both estimates of the distances—the 1 megaparsec of the American astronomers for Maffei 1 and the 2.7 megaparsecs of the French radio astronomers for Maffei 2—are correct and that the two galaxies appear close merely because of perspective. In that case, Maffei 1 would be at the outer limit of the local system and Maffei 2 would belong to that of Ursa Major–Camelopardus.

In this connection the radio observations announced by R. Love in January 1972 appear very significant. According to his observations, made with the half-mile interferometer at Cambridge, Maffei 2 is 4 megaparsecs distant and hence part of the M 81 system, but at a distance still greater than that found by the French astronomers. Moreover, while recording the distribution of neutral hydrogen, he found no traces of distortions attributable to Maffei 1, which therefore cannot be very near it. Thus it seems that the two objects are not connected.

But at this point the research has already entered a third phase, that of the search for and study of all other possible galaxies in the zone, however much obscured, which, as the discovery of the two Maffei galaxies has shown, can be found and studied, even if under inferior conditions. Infrared photographs of normal galaxies have been obtained at the Mount Wilson and Palomar Observatories in order to compare their appearance with that of highly obscured galaxies such as the two Maffei galaxies. Other astronomers have proposed, successively, that IC 342, then IC 10, and finally NGC 404 and NGC 1569 be included in the Maffei group.

After having advanced the hypothesis that the M 81 group is an enormous elongated complex, fully 10 million light-years across, which extends toward us, encompassing not only the Maffei galaxies and nearby IC 10 but even M 33 and the Andromeda galaxy, the French radio astronomers determined the distance of IC 10, which turned out to be 3 megaparsecs.

Meanwhile, in March 1972 L. Kohoutek and U. Haug, having conducted a new observational determination of the absorption, concluded that it is less than that assigned by the American astronomers and, hence, that the elliptical galaxy Maffei 1 is also about 3 Mpc distant. Thus both new galaxies would be outside the local system but near each other. According to the latest conclusions of the French radio astronomers, the large galaxy Maffei 1 and the four other smaller ones (IC 10, IC 342, NGC 1569, and Maffei 2) constitute an isolated group (which they call the "Maffei group") located between the group around M 81 and that in the vicinity of the Andromeda galaxy, M 31.

This last development in the research has shown, above all, that at the present time the most important consequence of the discovery of the two Maffei galaxies is a better knowledge of the nearby universe, just beyond the limits of the Galaxy. The concept of a local system may have to be totally revised, and it may not even exist in the sense attributed to it thus far, which limits it to being an extreme offshoot of the great Ursa Major–Camelopardus system.

E   LIFE IN THE UNIVERSE

The exploit of the two Viking space probes that landed on Mars in July and September 1976 has aroused worldwide enthusiasm on account of the perfection of the technology and for the scientific results achieved; at the same time it has caused one great disappointment. For thousands of years there has been talk of extraterrestrial life, and for more than a century it was believed that, if there were any planet on which some form of life might be sought with the certainty of finding it, that planet was Mars. The two Vikings were sent to Mars chiefly for this search, and the result has been, so far at least, negative. It is possible that a search carried out with other methods or in other regions of the planet may lead to a positive result in the near future. For the time being, the analyses and observations have given a result that comes as a cold shower for all those who support the theory of a plurality of inhabited worlds: that there can exist a planet with an atmosphere, climate, and soil capable of evolving; in short, a planet that is geologically alive and could harbor some form of life—on which nevertheless there is no life.

This result was expected for the moon, where the environment is absolutely inhospitable, but not for Mars, for which the major problem seemed to be not so much whether life exists there but what kind of life we might meet. Thus, after the negative result of the Vikings, further discussion of this question seemed in order, and one aspect of it in particular: the probability of finding inhabited planets in the universe.

Starting with considerations of statistics and probability, we may develop the inquiry as follows. As in the case of Mars, Man's interest was aroused above all by the thought of *how* there could be extraterrestrial life, but the fundamental problem, not yet solved, is *whether* life exists outside our earth and, if so, how frequently it occurs. An attempt has been made to calculate how many planets there are in our galaxy that support some form of life (not necessarily intelligent life). Their number can be estimated easily, at least in theory. It is given by the product of the number of stars in the Galaxy, the fraction of them that possess planetary systems similar to ours, the number of habitable planets in each system, and the probability that life does in fact develop on each planet providing favorable conditions. The result can be summed up in the following formula:

$$N = n \times f_p \times P \times p.$$

If we substitute for each letter in the formula the number expressing its most probable value (as far as we are able to determine it, with our present knowledge), we can deduce the probable number of planets in the Galaxy that harbor life.

The number of stars in the Galaxy is 10^{11}. The number of planetary systems is less certain. It can be derived by multiplying the number of stars in the Galaxy, n, by f_p—that is, the fraction of them that have a family of planets. In fact, we know of only one solitary planetary system in the whole universe, our own; but theories of the formation and evolution of the stars and the fact that most of the planets in our solar system possess satellites might lead us to think that nearly all stars should possess planets. On the other hand, we must exclude most double stars, where, except in the case of sufficiently separated pairs, it would be difficult for stable planetary systems such as ours to exist. In addition, O and B stars should also be excluded; they shine for so brief a period that no time is allowed for possible planets to evolve to the point where life could arise and develop. About half the stars are double. If we exclude the most widely separated pairs and include instead the O and B stars, we may conclude that f_p can be put equal to 0.5.

We can deduce the number of habitable planets, P, by considering our solar system, where there are at least two such planets (Earth and Mars) and another where the possibility is not completely excluded (Venus). Trying not to be too optimistic, let us take $P = 2$. Then, multiplying $2 \times 0.5 \times 10^{11}$, we get exactly 10^{11}—that is, a hundred billion habitable planets in the Galaxy. Note, however, that we have said *habitable,* not inhabited. To obtain this last quantity, we must further multiply by p, the probability that, when the conditions necessary for supporting life are found on a planet, life does in fact appear. This is the most uncertain factor in the whole problem.

If we assume that the number of habitable planets in a galaxy is 10^{11} and that the number of galaxies in the whole universe is 10^{10}, then the number of habitable planets in the universe is 10^{21}. Out of this incredibly great number of planets at least one (the earth) is inhabited: thus under the hypothesis (so pessimistic as to seem absurd) that ours is the only planet actually inhabited, the value of p would be 10^{-21}. Between this value and $p = 1$, which corresponds to the case when all the planets suitable for life are actually inhabited, there is an enormous range of possible values of p, and so far we have not succeeded in finding the right one.

The discovery of life on Mars could perhaps solve this problem. In fact, today it is unanimously agreed that any form of life on Mars could not come from Earth, because, even supposing that living spores could travel in space (as was suggested at the beginning of the present century), they would be sterilized by the very intense ultraviolet radiation coming from the sun. Thus if we were to find life on Mars, we should have to admit that, since the environmental conditions favorable to life developed on that planet independently of those on Earth, then life itself would have developed independently of terrestrial life, as if Mars were actually a planet of another system. In other words Mars can be considered as a planet chosen fortuitously among those provided with conditions favorable to life. If, then, some form of life were actually to be found there, it would indicate that the probability that life arises and develops whenever the possibility exists is very high, and p could be approaching or even equal to 1. In this case, the number of inhabited planets in the Galaxy would be about 100 billion and those in the whole universe would reach the incredible figure of 1 billion trillion.

Here, however, there is a very important point to be considered. If on a planet—for example, Mars—life proves to be completely absent, one might object that it has not yet developed but may someday. In fact, for any world there must be a period of time during which, although it is planetary in form, it has not yet developed to the stage where it would be suitable for the presence of life, just as there must be a period in which, for various reasons, the planet continues to exist but no longer harbors any form of life. One of these cases must occur sooner or later and can be attributed, for example, when the sun that illuminates and heats it increases or diminishes its temperature to such an extent that the temperature on the surface of the planet exceeds the limits within which life is possible, thus causing life to become extinct.

Our reasoning up to this point does not make much sense unless we formulate the question more precisely and, indeed, in a more interesting way: namely, how many planets in the Galaxy harbor some form of life at the present moment? To obtain this number we must multiply the result previously obtained for the formula by the time L during which a planet sustains life, expressed as a fraction of the total time that the planet exists.

If we suppose that the planet exists for 10^{10} years and that it contains living beings for at least half that time, the value of L will be 0.5.

When we consider the question from this point of view, we can no longer be so discouraged by the results obtained on Mars, for it may be that our exploration was made outside the favorable period. Let us then project from this result and calculate the number of planets in the Galaxy on which life presumably exists at the present moment, in two cases: first, assuming that $p = 1$ (that is, under the hypothesis that life appears on every planet sooner or later, whenever conditions favorable to life are formed); second, taking $p = 0.001$ (that is, assuming that even when a planet is adapted for life, life actually appears and develops in only one case in a thousand). In the first case we conclude that in the Galaxy at this moment life exists on 50 billion planets, and even in the second, more pessimistic case we still find a very large number of inhabited planets: 50 million.

Thus even in the less optimistic case the number of planets in the Galaxy that harbor life would be extremely high and in the whole universe would reach the incredible number of 500 million trillion.

This result applies to any form of life—vegetable, for instance—and does not necessarily imply the existence of intelligent humanoid beings. Such an event must be much rarer and in any case would be of much shorter duration. If we assume, with A. G. W. Cameron, that in all inhabited planets there exists a species that reaches a sufficiently high grade of intelligence and that this species takes 3,000 million years to develop and lasts 1 million years, the value of L would be 0.0003. The intelligent species present in the whole Galaxy at this moment would then be 30 million in the most favorable case ($p = 1$) and still 30,000 in the less favorable case ($p = 0.001$). In the latter case the number of planets in the whole universe inhabited by intelligent beings would be between 300 million million and 300,000 million million.

Thus even in the case when life would develop only in one in a thousand of those planets capable of harboring it, the number of inhabited planets in the universe would be extremely high, because of the huge number of galaxies, stars, and, almost certainly, planets.

Of course, this does not prove that the value of p could not be equal to 10^{-21}, which would mean that the earth is the only inhabited planet in the universe. It will never be possible to prove this, because that would involve exploring all the planets in the universe and always getting a negative answer. In other words, it is possible that one day we shall find life outside our earth (on a planet revolving around one of the nearer stars, say), but we shall never be able to prove the contrary, namely, that the earth is the only inhabited planet in the whole universe.

In conclusion we cannot at present say much more, but coming generations will certainly try to solve this fascinating problem by sending more probes to Mars and to other planets in the solar system, even by undertaking voyages with manned spaceships to the planets of the closer stars and by continuing to send out into space code signals that may someday bring a reply.

INDEX

Page numbers in italics refer to figures

Abell, G. O., 287, 292
Absolute magnitude, 119
  and diameter of galaxy, *262*
  of flare stars, 179–180
  of stars, 334, 336
Absorption, interstellar, 293
Acrylonitrile (CH₂CHCN), in interstellar
  space, 220
Adams, W. S., 108
AD Leonis (star), flares observed in, *177*
Adonis (minor planet), orbit of, *68, 70*
Aerobee missile, 267
Aitken, R. G., 103, 141
Alba Patera (Martian volcano), 56
Alcyone (star in Pleiades), 194
Aldebaran (star), size of, *124*
Algol (variable star), 137, 337, 341
Alpha Centauri (star), 99, 100, 110, 211
  distance from earth, 123
  orbit of, *104*
Altair (star)
  brightness of, 119
  distance from earth, 125
  size of, 125
Ambartsumian, V. A., 180, 185–186
Amino acids, in interstellar space, 220
Ammonia (NH₃), in interstellar space, 218
Ammonium hydrosulphite, on Jupiter, 74
Amor (minor planet), orbit of, *68*
Anderson, J. D., 74
Andromeda galaxy, globular clusters in,
  267–269
Andromeda nebula (M 31), 244–245,
  246–247, 248–252, 253, *355*
  distance from earth, 245–248, 291
  light curve for, 249
  and Maffei galaxies, 352
  mass of, 252
  novae of, 250
  radial velocity of, 296
  rotation of, 251
Antares (star), size of, *124*
Apollo (minor planet), orbit of, *68, 70*
Apollo 11, 328

Apollo 12, 19
Apparent magnitudes, 117. *See also* Ab-
  solute magnitude
Aquila (constellation)
  W 49 cloud in, 219
  variable stars in, 341
Arcturus (star), size of, *124*
Argelander, F. W. A., 340, 343
Argelander's catalog, 337
Arp, H. C., 250, 260, 261, 269, 282,
  284
Arp's diagram, *262*
Asiago Astrophysical Observatory,
  photographs from, *177, 196, 348, 349*
Associations, stellar, 347
A stars
  in gravitational contraction, 181
  size of, 122
Asteroid belt, 67, 68
Asteroids. *See also* Planets
  numbers of, 289
  research on, 70
Asterope (star in Pleiades), 194
Astronomical Society of the Pacific, 351
Astronomical unit (A.U.), 101
Astrophysical Journal, 307
Atkinson, R. d'E., 24
Atlases, celestial, 343–346. *See also*
  Catalogs
  Atlas Coeli Catalogue, 346
  Atlas of Peculiar Galaxies, 282
  Mount Palomar photographic, 257, 351
Atmosphere
  Jupiter's, 71, 74
  Martian, 48, 61, 63
  on minor planets, 69
  Neptune's, 79
  of planetary satellites, 85
  of Uranus, 78
  Venusian, 40, 41
Atomic energy, 3
Atoms, *28*
  characteristics of, 115–116
  splitting of, 30
ε Aurigae (star), 122, 135
ζ Aurigae (star), 135, *136*

Baade, W., 70, 158, 247, 250, 270
Baade's star
  light curve for, *159*
  pulsations of, 158
Babinet, J., 91
Background radiation, 323–324
Barnard's star, 113
  distance from earth, 125
  velocity of, 193
Becklin, E. E., 218
Becvar, A., 346
Bell Telephone Laboratories, 323
Berkeley, University of California at, 216
Bertola, F., 269, 286
Beryllium
  and expansion of universe, 323
  reactions of, *29*
Bessel, F. W., 108, 196, 333
Betelgeuse (star), size of, *124*
"Big Bang" theory, 324
Big Bear Solar Observatory, photo-
  graph from, *18*
Binary stars, 101–108
  eclipsing, 130, 132, 134
  spectroscopic, 131, *132–133,* 134
Black galaxy, 293
"Blink" method, of discovering variable
  stars, 337–338
Blue stars, 115
Bode's law, 67
Bond, W. C., 76
Bondi, H., 317, 318
Bonner Durchmusterung (BD), 337, 343
Booker, A. A., 143
Boron, reactions of, *29*
Bottinelli, L., 354
Bradley, E., 196
Bridges, between galaxies, 283, *284,*
  285, 292
Brightness
  of Cepheids, 145
  of flare star, 176
  measurement of, 117, 119
  of quasars, 304
  of star, 336
B stars, 120, 122

Burbidge, G. R., 270, 278, 280, 303
Burbidge, M., 278
Burnham, S. W., 103

Calcium, in interstellar space, 215
California Institute of Technology, 217
Callisto (satellite of Jupiter), 85
Cameron, A. G. W., 162
Canyons, Martian, 58, *59*
Capella (star), size of, *124*
Carbon monoxide (CO), in interstellar
  space, 219
Carbon-nitrogen cycle, *29,* 30
Carbon stars, 143
Carr, M. H., 58
Cassini, G. D., 71, 76
Castelli, Benedetto, 102
Catalog-index, of double stars, 105–106
Catalogs, celestial, 343–346
  Argelander's, 337
  in *Connaissance des temps,* 345
  General Catalogue of Variable Stars,
    143, 338, 341
  New General Catalogue, 345
  Pickering's, 337
  Pogson's, 337
Catania Observatory
  observations of, 178
  photograph from, *18*
Causality principle, 315
Celaeno (star in Pleiades), 194
α Centauri (double star), *104*
ω Centauri (cluster), 201
δ Cephei (giant star), 144, 145, 337
Cepheids, 144, 341
  brightness of, 145
  classical, 147
  in the Galaxy, 211
  in M 31, 245, 250
  in Magellanic Clouds, 237–239
  period-luminosity relation for, *146*
  RR Lyrae-type, 147
  variable characteristics of, *146*
Ceres (minor planet), 67
  orbit of, *68*
  size of, 69, 72

Certainty, concept of, 315, 316
Chromosphere
  description of, 32, *33*
  of sun, 11, 13, *14, 15*
  vortex structures of, *18*
Chryse Planitia (Mars), 59
Clouds
  Magellanic, 237–243
  of OH, 216–218
  Venusian, *42,* 43
Clusters. *See also* Globular clusters
  galactic, 192
  in intergalactic space, 294
  open, 192
Colombo, G., 34
Color index scale, *204*
Coma, 89
Coma Berenices (constellation), cluster
  of galaxies in, 285, *286,* 287
Comets, 87–98
  Halley's, 87, *88, 90*
  Ikeya-Seki, *88,* 93, *95*
  mass of, 91
  nucleus of, *92*
  numbers of, 289
  orbit of, *96*
  physical characteristics of, 89
  Tago-Sato-Kosaka, 91
  tail of, *92, 94–95*
Compact galaxies, 260–264, 294
Connaissance des temps, 345
Copernicus, N., 316, 333
Cordoba Durchmusterung, 346
Corona
  description of, 32, *33*
  of sun, 11, *12*
Cosmic rays, 19, *22–23,* 165
Cosmic vacuum, 243–244
"Cosmic void," 97
Cosmological principle, 317, 318, 319,
  321
Cosmology, 312
CP Lacertae, light curve of, *149*
Crab nebula, 154, *155*
  motions of expansion from, *156*
  pulsar of, *159*

  radio waves from, 157
Craters
  on Mars, 49
  on Mercury, 37, 39
Creation, continuous, 315–319, *321*
Crimean Astrophysical Observatory, 177
Cristaldi, S., *178*
Cyanogen (CN), in interstellar space, 219
61 Cygni (double star), 333
  distance from earth, 125
  orbits of, *104*
61 Cygni A, perturbed motion of, *112*
Cygnus (constellation), variable stars in,
  341

Dark zones
  nebulae as, 172
  on surface of sun, 5 (*see also* Sunspots)
Day, Mercurian, 39
Deimos (satellite of Mars), 66, 84, 85
Dembrowski, E., 103
de Mottoni y Palacios, G., 51
Dent, W. A., 306
de Vaucouleurs, G., 270, 287
D galaxy, 287
Displacement, measurement of, 193
Distances, determination of, 333–336
Dollfus, A., 76
Doppler effect, 130, 144, 296, 297, 303
Dorpat Observatory, 103
DQ Herculis (double star), 153
Draco (constellation), dwarf cluster in,
  257
Drake, F., 234
Dreyer, J. L. E., 345
Dust
  in intergalactic space, 293
  interstellar, 215
Dwarf galaxies, 254–260, 291
  in intergalactic space, 294
  in Sextant constellation, *259*
Dwarf stars, 122

Earth
  characteristics of, 47
  extraterrestrial view of, 190

Earth (*continued*)
  orbit of, *96*
  physical dimensions of, 1
  satellites of, 84
  size of, *36*
Eclipses
  solar, 11, *12, 14*
  of stars, 127–139
Eclipsing binary stars, 130, *132*, 134, 342
Eclipsing variables, 128. *See also* Variable stars
  in M 31, 250
Ecliptic, 35
Einstein, A., 313
Einstein effect, 302
Electra (star in Pleiades), 194
Electrons, 25
  and expansion of universe, 322
  in hydrogen, 212
Elementary particles, 25–31
Elements. *See also* Matter
  heavy, 31
  lightest, *28*
  origin of, 323
Eliasson, B., 218
Elliptical galaxies, 253, 254, 265, 266, 267
Elysium region, on Mars, 56
Encke's Comet, *88, 89*
Energy
  atomic, 3
  electrical, 3
  Fermi, 162
  and heavy elements, 31
  solar, 2–4, 24
Equatorial zones, on Mars, 63
Equinoxes, sun during, 32
o_2 Eridani B (white dwarf star), 109
Eros (minor planet)
  orbit of, *68*
  size of, 69
Eta Cassiopeia (star), 106
Europa (satellite of Jupiter), 85
Evolution
  cosmic, 319
  of galaxies, 285, 310

Evolutionary theories, 181
Ewen, H. I., 213
Excitation nebulae, *171*
Explosion
  great, 319–325
  stellar, 147–153

Fabricius, David, 337
Fechner, G. T., 117
Fermi, Enrico, 110
Fermi energy, 162
"Fermi gas," 110
Fernie, J. D., 143
Filaments
  solar, *15*
  on surface of sun, 13, *16–17*
Filter technique, for chromosphere observation, 13, 19
Fission, uranium atom, 30
Flares, solar, *18, 19–20, 23–24*
Flare stars, 173–180
  discovery of, 179
  H α emission line of, 180
  observation of, *177*
"Forbidden lines," of spectrum, 169
Ford, W. K., 251
Formaldehyde (H_2CO), in interstellar space, 219
Formamide (NH_2COH), in interstellar space, 219
Formax (constellation), dwarf cluster in, 257
Formic acid (H_2CO_2), in interstellar space, 219
Fraunhofer Institut, photographs of, 22
F stars, size of, 122
Fusion processes, 31

Galactic cluster, 192
Galaxies
  barred spiral (SB), *258–259*
  black, 293
  bridges between, 283, *284*, 285, 292
  chains of, *284*, 285
  classical, 263

Galaxies (*continued*)
  clusters of, 285
  compact, 260–264, 294
  D, 287
  distances of, 297, 299
  double, *282,* 283
  dwarf, 254–260, 291, 294
  elliptical, 253, 254, 265, 266, 267
  evolution of, 285, 310
  flight of, 316
  globular clusters around, 267
  groups of, 281–291
  IC 10, 356
  IC 342, 356
  IC 1805, 348, 349
  infrared emission of, 281
  interacting, 285
  irregular, 253, 254
  local system of, 287, *352,* 357
  M 82, 271–272, *273,* 274, 280, 324, 354
  M 87, *268*
  M 101, *255*
  Maffei, 347–348, *349–350,* 351, *352,*
    353–357
  movement of, 296–301, 311
  multiple, 283, *284*
  NGC 404, 356
  NGC 891, *256*
  NGC 1275, 278, *279,* 280
  NGC 1569, 356
  NGC 4438, *266*
  NGC 5128, 303
  NGC 5432, 282
  NGC 5435, *282*
  NGC 7320, *284*
  numbers of, 252, 289
  quasi-stellar (QSG), 307
  red shifts of, 297, *298*
  SB, 253
  Seyfert, 274–278, *279,* 280–281
  small, 292
  spectra of, 297
  spiral (S), *258–259,* 277, 282, *288*
  Ursa Major-Camelopardus system of,
    354, *355*
  variety of, 253–254

  violent events in, 277
Galaxy, the, 207
  center of, 229–236
  central zone of, *232*
  globular clusters in, 267–269
  hydrogen in, 231
  mass of, 226–229
  motion of, 221–226
  nucleus of, 233
  number of stars in, 229
  size of, 211
  structure of, 211–213
Galileo, 76, 84, 102, 328, 333
Galle, J. G., 79
Gamow, G., 323
Ganymede (satellite of Jupiter), 85, *86*
Gaposchkin, S., 139
Gas, in intergalactic space, 293, 294.
    See also Elements
Gasteyer, C., 110
Geisel, G. L., 151
General Catalogue of Variable Stars,
    143, 338, 341
Geocentrism, 206, 299
Geological eras, and earth displacement,
    223–224
Geometry, Euclidean, 313
Giant stars, 122, 126, 203
Globular clusters, *200,* 201, 202
  in Andromeda galaxy, 251
  around galaxies, 267
  in the Galaxy, 211
  on H-R diagram, 203, *204*
  in intergalactic space, 292
  in Large Magellanic Cloud, 239
  orbits of, 226
  47 Tucanae, *238*
Gold, T., 317, 318
Goodricke, J., 137, 337
Granules, on surface of sun, 5, *6,* 7
Gratton, L., 310
Gravitation, 186, 312
"Greenhouse effect," on Venus, 43
Grueff, Gavril, 351
G stars, size of, 122
Guérin, P., 76

Hall, A., 103
Halley, E., 89, 201, 343
Halley's comet, 87
  orbit of, *88*
  photographs of, *90*
Halo stars, 226
Hardie, R. H., 81
Haro, G., 179, 182, 185
Hartmann, J., 215
Harvard College Observatory, photo-
  graphs from, *12, 238*
Harvard University, 216
Haug, U., 357
HD 47129 (Plaskett's star), 107
Helium, 31, 323. *See also* Elements
Hellas basin, on Mars, 58
Heraclitus, 296
Herbig, G. H., 182
Herbig-Haro objects, 182, *184,* 188
Hercules (constellation), 198
  cluster of galaxies in, *288*
Herculis (nova), *149*
α Herculis (star), size of, *124*
Hermes (minor planet), orbit of, 70
Herschel, W., 77, 102, 103, 201, 234,
  316, 328
Hertzsprung, E., 120
Hertzsprung-Russell (H-R) diagram,
  120, *121,* 122
  for globular clusters, 203
  for M3, *204*
  for nearby stars, *124,* 125
  for Orion association, 182, 183
  for Pleiades, *204*
  for stellar evolution, 181
Heslod, 345
Hidalgo (minor planet), orbit of, *68,* 70
High Altitude Observatory, photographs
  from, *12*
High-velocity stars, 224, *225,* 226
Hipparchus, 148, 343, 345
Hodge, P. W., 239, 300
Hoffmeister, C., 293, 294
Holden, E. S., 103
Homer, 345
Honda, M., 151

Horse's Head (nebula), 172
Houtermans, G. F., 24
Hoyle, F., 303, 318
H-R diagram. *See* Hertzsprung-Russell
  diagram
Hubbard, W. B., 74, 254
Hubble, E. P., 250, 258, 278
Hubble constant, 297, 299
Hubble's law, 297, 299, 304
Humason, M., 80, 297
Hussey, W. J., 103
Huyghens, C., 75
Hyades (cluster), 194
Hydrocyanic acid (HCN), in interstellar
  space, 219
Hydrogen
  characteristics of, 26
  emission line of, 13, 15, 169
  and expansion of universe, 323
  galactic neutral, 212–213
  heavy, 26, 31
Hydrogen bomb, 24, 30
Hyland, A. R., 151

IC 10 (galaxy), 356
IC 342 (galaxy), 356
IC 1805, 348, 349
Icarus (minor planet), orbit of, *68,* 70
Ikeya-Seki comet, *88,* 93, 95
Indeterminacy principle, 315
Index Catalogue, 345
Infrared, photographing in, 347
Infrared emission, 280–281
Infrared radiation
  discovery of, 234
  of quasars, 309
Infrared source, at center of Galaxy,
  234–235
Innes, R. T. A., 103
Intergalactic bridges, 283, *284,* 285, 292
Intergalactic space, 291–295
Interstellar dust, 215
Interstellar space, 25, 215–226
  density of, 172
  neutral hydrogen in, 215–216
  OH radical in, 216

Io (satellite of Jupiter), 85
Irregular galaxies, 253, 254
Island formations, on Mars, *60*
"Island universes," 252
Isocyanic acid (HC₃N), in interstellar
    space, 219
Isotopes, 26

Janssen, J., *7*
Jonckheere, R., 103
Joy, A. H., 141
Juno (minor planet), 67, 69
Jupiter
    atmosphere of, 74
    characteristics of, 71–75
    internal temperature of, 75
    satellites of, 84
    size of, 36, 71, 72
    surface of, 74

Kelvin, W. T., 21, 24, 31
Keplerian motion, *222*
Kepler's second law, 226
Kepler's third law, 103, 106, 130, 227
Kiloparsec, 334
Kinman, T. D., 308
Kitt Peak National Observatory, photo-
    graphs from, *159, 279*
Kleinmann, D. E., 151
Kohoutek, L., 357
K stars, size of, 122
Kuiper, G. P., 80, 81
Kukarkin, B. V., 338

L 726-8 (star), 173, 176
Lacchini, G. B., *174*
Lamla, E., 308
Landau, Robert, 351
Large Cloud. *See* Magellanic Clouds
Leavitt, H. S., 237
Leighton, R., 218
Leo (constellation), dwarf cluster in, 257
Le Verrier, U., 79
Leyden Observatory, minor planet re-
    search of, 69
LGM ("Little Green Men"), 157

Lick Observatory, 182
    photographs from, *174–175, 184, 282,
    284*
    telescope, 282
Life, in universe, 25, 358, 362
Light curve
    for Baade's star, *159*
    of β Lyrae, 138
    for M 31, 249
    of Mira Ceti, 142
    of novae, *149*
    for stellar brightness, 128–129
    of 3C 446, 308
    of T Orionis, *174–175*
Light-year, 333
Liller, W., *88*
Limb-darkening effect, of stellar eclipses,
    129
Lithium, 26, 27. *See also* Elements
    and expansion of universe, 323
    reactions of, *29*
Local system, of galaxies, 287, *352, 357*
Loiano astronomical station, photo-
    graphs from, *94–95*
Love, R., 356
Low, F. J., 151
Lowell, P., 79
Lowell Observatory, photographs from,
    50, 82–83
Low-velocity stars, 224, *225*
LP 321–98, mass of, 109
Lunae Planum (Mars), 59
Lundmark, K., 154
Luyten, W. J., 108, 173
Luyten 726–8 B (UV Ceti), 107–108, 176,
    179
Lynds, R. C., 272
β Lyrae, 137, 337
    components of, *139*
    distance from earth, 139
    light curve of, *138*
    in M 31, 250
β Lyrae system, model of, *142*

M 3, H-R diagram for, *204*
M 13 (NGC 6205), *200*, 201

M 20 (nebula), *171*
M 31. *See* Andromeda
M 32 (NGC 221), 252
M 42 (nebula), 167, 172, *174–175,* 179, 180
M 51 (galaxy), 281
M 67 (cluster), *197*
M 81, 354, *355,* 356
M 82 (galaxy), 271–272, *273,* 274, 280, 324, 354
M 84 (galaxy), *266*
M 86 (galaxy), *266*
M 87 (galaxy), 265, 266, 267, *268,* 280, 285
  central jet of, 269–271
  distance from earth, 269
  synchrotron radiation of, 270
  x-ray emission from, 267
M 101 (galaxy), 255
Maffei 1 (galaxy), 347, *349, 350,* 353, 354, *355, 356*
Maffei 2 (galaxy), 347, *349, 352,* 353, 354, *355, 356*
"Maffei group," 357
Magellanic Clouds, 237–243, 252, 253, 260
Magnetic field
  Jupiter's, 74
  on Mercury, 37
  solar, 18, 19
  of sun, 13
Maia (star in Pleiades), 194
Main sequence, 121, 181. *See also* Hertzsprung-Russell diagram
  for Orion association, *183*
Marine life, and cosmic rays, 165
Mariner 2, Venus observations of, 67
Mariner 4, discoveries by, 67
Mariner 6, photographs from, 49, *51*
Mariner 7, photographs from, 49, *51*
Mariner 9
  Martian observations of, 58
  photographs from, *52–53, 54, 55, 57, 62, 85, 86*
Mariner 10
  observations of, 34, 37

  photographs from, *38, 42*
  Venus observations of, 41
Mars
  atmosphere of, 61–63
  canyons of, 58, *59*
  characteristics of, 47–67
  furrows on surface, 58, *59*
  map of, *52–53*
  polar cap of, 48, *50–51, 54, 55,* 61–63
  satellites of, 66, 84, 85
  size of, *36,* 47
  south pole of, *62*
  surface of, 49, *51,* 63, *64–65,* 66
  and Viking space probes, 358
  volcanoes of, 56, *57,* 58
Marsden, B. G., 93
Mass, determination of, 26
Massachusetts Institute of Technology, 216
Mass-luminosity ratio, for compact galaxies, 261
Matter. *See also* Elements
  composition of, 25
  creation of, 318–319
  degenerate state of, 109–110
  of neutron stars, 162
  protomatter, 186
Méchain, P., 345
Megaparsec, 334
Mercury
  characteristics of, 34–39
  orbit of, *35*
  size of, *36, 86*
  surface of, 37, *38*
  temperatures on, 37
Merope (star in Pleiades), 194
Messier, C., 153, *155,* 167, 201, 345
Meteor shower, *96*
Methyl alcohol (CH_3OH), in interstellar space, 219
Methylamine (CH_3NH_2), in interstellar space, 219
MGC 457 (cluster), 194
Micrometeorites, 67
Milky Way, 99. *See also* Galaxy, the
Minkowski, R. L., 185, 278

Mira (star), 337, 341
Mira Ceti (pulsating red star), 140, *142*
Mira-type variables, 341
Mizar (star), 102
  spectra of, *132–133*
Molecules. *See also* Elements in in-
    terstellar space, 219, 220
  in interstellar space, 219, 220
  polyatomic organic, 219
Mons Elysium, 56
Mons Olympus, 56, *57,* 58
Montanari, Gemignano, 337
Moon, earth's, 84
  astronomical observatories on, 329
  size of, 72, *86*
  sun viewed from, 2
Moons, Martian, 66
Morgan, W. W., 212, 287
Morikura astronomical station, photo-
    graphs from, *95*
Mount Palomar observatory
  minor planet research of, 69
  photographs from, *14, 15, 72, 132–133,
    150, 155, 168, 170, 197, 208–209,
    246–247, 255, 256, 258–259, 262,
    266, 273, 279, 282, 284, 286, 288,
    298, 305, 350*
  Schmidt telescope at, 261, 283, 292
  spectroscopic measurements of, 302
  sunspot observations of, *10*
  200-inch reflector of, 351, 352
Mount Wilson Observatory, 173
  photographs from, *8, 14, 15, 72, 90,
    132–133, 150, 155, 168, 170, 197,
    208–209, 246–247, 255, 256, 258–
    259, 262, 266, 273, 279, 282, 284,
    288, 298, 305*
  sunspot observations of, *10*
Mrkos (comet), *94–95*
M stars, 120
  in gravitational contraction, 181
  size of, 122

Nançay radiotelescope, 354
Nebulae
  Andromeda, 244–245, *246-247,* 248, 252

composition of, 169
crab, 154, *155, 156,* 157, *159*
as dark zones, 172
determining distances of, 154
excitation, *171*
M 42, 172
NGC 1977, 188
NGC 2023, 190
NGC 2024, 190
Orion, *168,* 249
reflecting, 194
rosetta, *170*
Tarantula, 240, *241, 242,* 294
Trifid, *171*
Neptune
  characteristics of, 79
  discovery of, 79–80, 113
  satellites of, 85
  size of, *36*
Neugebauer, G., 151, 218, 294
Neutral hydrogen, 212–213
  in Andromeda galaxy, 251
  distribution of, *214*
  in the Galaxy, 231
  in interstellar space, 215–216
Neutrinos, and expansion of universe,
    322
Neutrons, 25, 322
Neutron stars, 157, 158
  characteristics of, 160, *161*
  matter of, 162
New General Catalogue, 345
Newton, I., 186, 312, 328
NGC 205, 252
NGC 221 (M 32), 252
NGC 404 (galaxy), 356
NGC 891 (galaxy), *256*
NGC 1275 (galaxy), 278, *279,* 280
NGC 1569 (galaxy), 356
NGC 1977 (nebula), 188
NGC 2023 (nebula), 190
NGC 2024 (nebula), 190
NGC 4438 (galaxy), *266*
NGC 5128 (galaxy), 303
NGC 5432 (galaxy), *282*
NGC 5435 (galaxy), *282*

NGC 7320 (galaxy), *284*
Nix Olympica (Mars), 56
Noachis region, of Mars, 58
Nomenclature, for variable stars, 340–341
Novae
  characteristics of, 148, *150,* 151
  double star characteristics of, 153
  light curve of, *149*
  in M 31, 245, 250
  ultraviolet observations of, 152
Nova Herculis, *149*
Nova Persei, *150*
Nova Serpentis, 151, 152
Nuclear bombs, 27
Nuclear reactions, *28*
Nucleus, of sun, *33*

O association, 180
Observatoire de Meudon, photographs from, *16, 17*
OH 471 (quasar), distance of, 304
OH radical, in interstellar space, 216–218
Olbers, W., 67, 312
Oort, J. H., 354
Open cluster, 192, *204*
Ophiuchus (constellation), variable stars in, 341
OQ 172 (quasar), 304, *324*
Orbits. *See also* Rotation
  of binary stars, 128
  of comets, 87, *88*
  of high-velocity stars, 226
  of minor planets, 70
Organic molecules, in outer space, 220
Origin, of universe, 322
Orion association, H-R diagram for, 182
Orion (constellation), 167
ϑ Orionis, 167, 187, 188
ζ Orionis, 190
Orion nebula, *168,* 240
  flare stars in, 179
  water vapor in, 219
O stars, 120, 122

Pallas (minor planet), 67, 69, 72

Parallax method, of determining stellar distance, *335*
Parenago, P. P., 182, 338
Parsec, 334
Particle physics, 25–31
Peery, B. F., 123
β Pegasi (star), *124*
Penumbra, of sunspots, *8,* 9
Period-luminosity relation, for classical Cepheids, *146,* 147
Perseus (cluster), 194, *196*
Perseus A (radio source), 278, *279*
Phase effect, of stellar eclipses, 129
Phobos (satellite of Mars), 66, 84, 85, *86*
Photometric binary star, 128
Photons, and expansion of universe, 322
Photosphere, 5, 9, 11, 21, *33*
Physics, 326
Pickering's catalog, 337
Piggott, E., 337
Pine Torinease Observatory, photographs from, *149*
Pioneer 10, Jupiter observations of, 74, 75
Pioneer 11, Jupiter observations of, 74
Planets, 25
  existence of, 113–114
  in the Galaxy, 358
  giant, 71–84
  habitable, 359
  minor, 67–71
  of nearest stars, 111–114
  numbers of, 289
  orbits of, 34, *35*
  satellites of, 84
  size of, *36*
Pleiades (cluster), 192–194, *195*
  distance from earth, 194
  H-R diagram for, *204*
  proper motions of, *196*
Pliny, 148, 343
Pluto, 82
  characteristics of, 79–84
  mass of, 80
Pogson's catalog, 337
Polar caps, Martian, 48, 50–51, *54, 55,*

Polar caps, Martian (*continued*)
  61–63
Population I stars, 203, 205, 342
Population II stars, 203, 205, 226, 233,
  341
Pores, on surface of sun, 5, 7
"Positive on negative" method of discov-
  ering variable stars, 337–338
Prague International Congress, 275
Prasepe (cluster), 194
Pressure. *See also* Atmosphere
  in central zone of sun, 21
  on Venus, 43
Probability, concept of, 315
Procyon (star), 125
Procyon B (white dwarf), 109
Prominence-filament, evolution of, *16–17*
Prominences, solar, 13, *14, 15, 16–17*
Proper motions
  measurement of, 193
  of Pleiades, 196
Protomatter, 186
Protons, 25
  and expansion of universe, 322
  in hydrogen, 212
Proxima Centauri (flare star), 110, *112,*176
Ptolemy, Almagest of, 343
Pulkovo Observatory, 103, 105
Pulsars
  discovery of, 157
  optical, 158
Pulsating stars, 140
  white, 143–147
Purcell, E. M., 213

Quantum theory, 162
Quasars, 301–310
  distance from earth, 304
  evolution of, 310
  OH 471, 304
  OQ 172, 304, 324
  structure of, 309
  variability of, *308*
Quasi-stellar galaxies (QSG), 307

R 76 (star), 240

Radiation
  background, 323–324
  infrared, 234, 309
  synchrotron, 270
  ultraviolet, 309
Radiative zone, of sun, *33*
Radio astronomy, 213
Radio emissions
  in the Galaxy, 234, 235
  from Maffei galaxies, 356
  from quasars, 302
Radio sources
  diminishing of, 324
  galaxies as, 280
  Perseus A, 278, *279*
  quasi-stellar, 302
Radio telescopes, 324
Raimond, E., 218
Rays, solar, 19, *22–23*
Red dwarf stars, 126, 179
Red shift
  interpretations of, 302–304
  in spectra of galaxies, *298*
Red spot, Jupiter's, 71, *72,* 73
Red stars, 115, 122
  pulsating, 140–143
Reflection nebulae, *171*
Relativity theory, 27, 313
River bed formations, on Mars, *60*
Rodono, M., *178*
Rosetta nebula, 170
Rosino, L., 179, 250
Rotation. *See also* Orbits
  of Andromeda nebula, 251
  galactic, *222*
  of the Galaxy, *225*
  of Mars, 48
  of Pluto, 81
  of Saturn, 75
  of sunspots, 9, *10*
  of Uranus, 77–78
RR Lyrae-type Cepheids, 147
RR Lyrae-type variables, 211
Rubin, V. C., 251
Russell, A. A., 120
RW Tauri (star), 135, *136*

Sacramento Peak Observatory, photographs from, *14–15*
Sagittarius (constellation), variable stars in, 341
Sagittarius A, 234–235
Sagittarius B 2 cloud, water vapor in, 219
Sandage, A., 269, 272, 305, 307, 309, 322
Sandstorms, Martian, 58, 61
Satellites. *See also* Moon
  numbers of, 289
  planetary, 84–87
Saturn
  characteristics of, 75–77
  rings of, 75–77, *82–83*
  satellites of, 85
  size of, *36*
Sawyer, H., 340
Schiaparelli, G. V., 34, 56, 103
Schmidt, B., 344
Schmidt, M., 307
Schmidt telescope, 170, 247, 261, 266, 286, 290, 292, 348, 349
Schönfeld, E., 343
Schwassmann-Wachmann comet, orbit of, *88*
Sculptor (constellation), dwarf cluster in, 257
Searle, L., 294
Secchi, A., 13
Serpentis (nova), 151, 152
Setti, G., 303
Sextant (constellation), dwarf galaxy in, *259*
Sexton, J. A., 239
Seyfert, C. K., 275, 278
Seyfert galaxies, 274–278, *279,* 280–281 307, 309
Shapley, H., 254, 257, 285
Shklovskii, I. S., 163, 218, 270
Sirius (star), 108
  apparent motion of, *112*
  brightness of, 119
  distance from earth, 125
  size of, *124, 125*

spectrum of, 117
Sirius B (white dwarf), 108
  discovery of, 113
  mass of, 109
  size of, *124*
Sky
  Jupiter, 87
  Martian, 66
  Plutonian, 81
  Saturn, 87
  Uranian, 78
Slipher, V. M., 296
Small Cloud. *See* Magellanic Clouds
Smith, H. J., 305
Solar system, inner and outer regions of, 35. *See also* Planets; Sun
Space. *See also* Time
  curved, 314
  intergalactic, 291–295
  interstellar, 25, 172, 215–226
  limits of, 325
  mean density of, 295
  unbounded, 314
Spectra, of stars, 114, *118*
Spectroscope, 114
Spectroscopic binaries, 131, *132–133,* 134
Speed, of earth in space, 1
Spinrad, H., 351
Spiral galaxies, 253
Sproul Observatory, 113
Stars, 25. *See also* Galaxies
  absolute magnitude of, 334, 336
  birthplace of, 180–187
  blue, 115
  carbon, 143
  cloud of, *208*
  colors of, 115
  densities of, 123
  determination of distances of, 333–336
  diameters of, 122
  double, 101–108 (*see also* Binary stars)
  eclipses of, 127–139
  flare, 173–180, 342
  in globular clusters, 202–203
  high-velocity, 224, *225,* 226

Stars (continued)
in intergalactic space, 294
low-velocity, 224, 225
multiple, 110
neutron, 157, 158, 160, 161, 162
numbers of, 289
nurseries for, 186
populations I and II, 203
pulsating, 140, 143–147
size of, 124
spectra of, 114, 118
types of, 114–123
variable, 337–342
velocities of, 225
weight of, 106–107
white dwarfs, 108–110
white pulsating, 143–147
yellow, 115
Stefan quintet, 284, 285
Stein (star), distance from earth, 125
Stellar explosions, 147–153
Strand, K. A., 186
Stratoscope observations, 8
Struve, O., 103, 137
Struve, W., 103
Sulphuric acid, in Venusian atmosphere, 43, 44
Sun
apparent size of, 83
appearance of, 4–5, 6–8, 9, 10, 11, 12, 13, 14–18, 19–20
central temperatures of, 21
and cometary activity, 91, 93, 95
corona of, 11, 12
description of, 32, 33
eclipse of, 11
energy derived from, 2–4, 24
interior of, 20–25
motion of, 223
origin of heat inside, 27
size of, 124
solar flares of, 18, 19–20, 23–24
Sunspots, 5, 6, 8
activity of, 9
periodic variations in, 20, 22–23
Superclusters, 287

Supernovae, 153–166
cosmic ray dose, 164–166
listing of, 341
radio emissions from, 157
types of, 163
Synchrotron radiation, 270

Tago-Sato-Kosaka (comet), 91
Tamman, G. A., 322
Tarantula (nebula), 240, 241, 242, 294
T association, 180
Taygete (star in Pleiades), 194
Temperature, in central zone of sun, 21
Temperature, surface
on Mars, 63
Mercury's, 37, 39
of sun, 4
on Venus, 43
Terrell, 303
Terry, K. D., 164
Tharsis region, on Mars, 49
volcanoes of, 56
Thermonuclear reactions, and gravitational contraction, 181–183
3C 273 (quasar), 304, 305, 306
3C 446 (quasar), 306, 308
Time. See also Space
cosmic, 319
limits of, 325
Titan (satellite of Saturn), 85, 86
Titania (satellite of Uranus), 86
Titanium oxide, in atmosphere of star, 116
Tombaugh, Clyde, 79, 80
T Orionis (flare star), light curve of, 174–175
Trapezium (star cluster), 240, 242
Trifid nebula, 171
Trimble, V., 156
Tritium, 26
Triton (satellite of Neptune), 85, 86
Tsuruta, S., 162
47 Tucanae (globular cluster), 238
Tucker, W. H., 164
Tucson Observatory, 173

Uhuru (artificial satellite), 234, 278
Ultraviolet radiation, of quasars, 309
Umbra, of sunspots, 9, 13
Universe
  evolution of, 312
  expansion of, *200, 320*
  life in, 358–362
  limits of, 314
  static, 317–318
  steady-state, 318–319
  theories about, 316
  uniformity of, 317
  volume of, 323
Uranium
  characteristics of, 26
  fission of, 30
Uranium bomb, 30
Uranus
  atmosphere of, 78
  characteristics of, 77–79
  distance from sun, 97
  satellites of, 85
  size of, *36*
Ursa Major (constellation), M 82 galaxy
  in, *273*
Ursa Major-Camelopardus system, 354,
  355, 356, 357
Ursa Minor (constellation), dwarf cluster
  in, 257
UV Ceti (flare star), 107–108, 176, 179
UX Monocerotis (star), 135

Valles Marineris (Mars), 58
van de Hulst, H., 212, 213
van de Kamp, P., 113
van den Bergh, 287, 353
van den Bos, W. H., 103
van Maanen, A., 173
Variable stars, 337–342
  discovery of, 337–*339*
  long-period, 143–144
  nebular, 342
  nomenclature for, 340–341
  "suspected," 340
Vega (star), *124*
Velocities, escape, 277

Venera 8, Venus observations of, 41, 46
Venera 9, photographs from, 44, *45*
Venera 10
  photographs from, 44, *45*
  Venus observations of, 46
Venus
  characteristics of, 40–46
  clouds of, *42, 43*
  Russian landings on, 41
  size of, *36,* 40
  surface of, 44, *45, 46*
Vesta (minor planet), 62, 69
Vetesnik, M., 251
Viking space probes, 61, *64, 65,* 358
Virgo (constellation), cluster of galaxies
  in, 285
Volcanoes, Martian, 56, *57*
von Helmholtz, H. L., 21, 24, 31
Vorontsov-Velyaminov, B., 285
VV Cephei (star), 122–123, 135

W3 area, 217
W 49 cloud in Aquila, water vapor in, 219
Walker, M. F., 81, 153, 182, *183,* 275
Walpole, H., 347
Water, possibility of, 39
Water vapor, in interstellar space, 218–
  219
Westerlund, B. E., 241, 242
Whipple's model, of comet mass, *92*
White dwarfs, 108–110, 157
"White nights," 199
White stars, 115
Wien's law, 115
Wilson Observatory. *See* Mount Wilson
  Observatory
Wind, on Mars, 63
Wirtanen, C. A., 308
Wolf, M., 69
Wolf 630 (star), 106
Wolf number, 22
Woltjer, L., 303

X-ray emission
  from M 87, 267
  of NGC 1275, 278

Year
  Martian, 48
  Neptunian, 79
Yellow stars, 115
Yerkes Observatory, 212
YY Gem (star), size of, *124*
YZ C Mi, flare of, *178*

Zero-age line
  on H-R diagram, 181
  for Orion association, *183*
Zodiacal light, 32
Zwicky, F., 157, 261, 283, 287, 289, 290,
    291, 293, 294